DILUTED
MAGNETIC
SEMICONDUCTOR
NANOSTRUCTURES

**A BOOK BASED ON SEVEN YEARS OF RESEARCH WORK
IN THE FIELD OF NANOSCIENCE AND NANOTECHNOLOGY**

DURGA PRASAD GOGOI

INDIA · SINGAPORE · MALAYSIA

Notion Press

Old No. 38, New No. 6
McNichols Road, Chetpet
Chennai - 600 031

First Published by Notion Press 2019
Copyright © Durga Prasad Gogoi 2019
All Rights Reserved

ISBN 978-1-64678-986-3

DEDICATED TO

This BOOK is dedicated to my wonderful parents Late Sukhen Chandra Gogoi and Late Panchami Gogoi who formed part of my vision and taught me the good things that really matter in life. I am grateful for my Wife Ms. Chenehi Gogoi, Daughter Miss Epshita Manashi (Maina) and Son Mr. Abhinandanan Manash (Bhaity) for offering me the best co-operation from their ends; I am glad to be one of them. Finally, I would like to dedicate this to my elder sister Ms. Roti Prova Gogoi, whose inspiration and cooperation brought me to this end.

Contents

Preface 7
Abbreviations 9

Chapter 1. Introduction 11
    1.1 Quantum confinement effect 12
    1.2 Transition metals 15
    1.3 Semiconductor Nanostructured materials 16
    1.4 Diluted magnetic semiconductors (DMS) 18
    1.5 II-VI based DMS 19
    1.6 Oxide based DMS 21
    1.7 ZnO based DMS 22
    1.8 ZnS based DMS 25
    1.9 Spintronics 27
    1.10 Objectives of the present study: 35

Chapter 2. Fabrication of DMS nanostructures 37
    2.1 Materials 38
    2.2 Fabrication of Bare ZnO nanostructures 39
    2.3 Fabrication of TM doped ZnO nanostructures 40
    Conclusions 42

Chapter 3. Experimental techniques 43
    3.1 X-ray Diffraction 44
    3.2 UV-Visible spectroscopy 47
    3.3 Photoluminescence spectroscopy 48
    3.4 Transmission Electron Microscope (TEM) 51
    3.5 Scanning Probe Microscope (SPM) 53
    3.6 Electron Dispersive X-ray spectrometry (EDX) 55
    3.7 Fourier Transform infrared spectroscopy (FTIR) 55

3.8   Superconducting quantum interference device (SQUID) 56

3.9   Significant observations 57

Chapter 4.   ZnO based DMS nanostructures 59

4.1   Mn doped ZnO 59

4.2   Co doped ZnO 79

4.3   Ni doped ZnO 89

4.4   Significant observations 98

Chapter 5.   ZnS based DMS nanostructures 101

5.1.   Mn doped ZnS 102

5.2   Cr doped ZnS 106

5.3   Significant observations 113

Chapter 6.   Magnetic measurements 115

6.1   Magnetic measurement of bare ZnO 115

6.2   Magnetic measurement of TM doped ZnO 116

6.3   Magnetic measurement of TM doped ZnS 127

6.4   Significant observations 131

6.5   Possible applications in spintronic devices 133

6.6   Possible applications in Luminescent devices 137

6.7   Limitations 140

6.8   Future directions 141

Appendix 143

Paper Presented in Seminar/Conference by the Author 145

Papers/Chapters Published by the Author 147

References 149

Preface

This book is based on my seven years of research experience in the field of nanoscience and nanotechnology. The book "Diluted magnetic semiconductor nanostructures" is meant for new researchers and students in this area. This book represents my original work in the field of diluted magnetic semiconductor nanostructures starting from chemical synthesis to few characterizations. Chemical synthesis is an inexpensive and simple route to synthesis nanostructures. In this book, the low cost, user friendly techniques to fabricate ZnO based nanostructures have been discussed. It will certainly help students, researchers and concerned people in this area. Characterization of structural, optical and magnetic properties of the nanostructures by using High Resolution Transmission Electron Microscopy (HRTEM), Atomic Force Microscopy (AFM), Magnetic Force Microscopy (MFM), Electron Dispersive X-ray (EDX), X-ray Diffraction (XRD), UV-Visible spectroscopy (UV), Photoluminescence study (PL), Fourier Transform Infra-red Spectroscopy (FTIR), Super Conducting Interference Device (SQUID) etc. have been discussed in the book.

As of today, we notice an ever increasing trend of research in nanoscience and nanotechnology. Nanotechnology is considered to be the future miniaturization technology with wide variety of potential applications. Nanotechnology concerns with the study of ultrafine material structures with average crystalline size of the order less than several nanometers. At this quantum-realm scale, quantum mechanical effect is important. Nanostructured materials exhibit unique characteristics due to their existence in the transition region between bulk solid and molecular structure. Different nanostructured materials with variable dimensions but having similar chemical composition may behave differently.

The book is divided into 6 chapters, each of which again splits into subsections. The first chapter is introductory in nature and the background of semiconductor nanostructures, II-VI based semiconductors, spintronics etc. is described in this chapter. Chapter-2 describes the synthesis details through inexpensive chemical and solid state reaction routes. Chapter -3 deals in different characterization techniques. Chapter-4 deals in the fabrication and characterization of bare ZnO and transition metal doped ZnO nanostructures. Different aspects from my investigations on ZnS based DMS nanostructures are highlighted in chapter-5. Finally, Chapter-6 is focused on magnetic measurements and observations.

I am grateful to Prof. Amarjyoti Choudhury, ex. Pro. Vice chancellor, Tezpur University, Prof. Gazi Amin Ahmed and Prof. Dambarudhar Mohanta of Tezpur University who supervised and guided me in the whole work. I am also thankful to G. A. Stanciu, Centre for Microscopy-Microanalysis & image Processing, University Politehnica" of Bucharest, Romania, for offering AFM and MFM facility. I am also very much grateful to Prof. Alok Banerjee, UGC, DAE consortium, Indore for the opportunity to utilize the facility of advanced scientific tool 'SQUID'. I am also very much grateful to the people at SAIF centre, NEHU, Shillong, for offering me HRTEM facility. I am indebted to many friends for assistance and help at all stages.

Finally, I am thankful to the publisher group of "Notion press" for their keen interest in bringing out this book.

Dr. Durga Prasad Gogoi

Abbreviations

AFM	Atomic force microscopy
CCD	Charged couple device
CVD	Chemical vapour deposition
DMS	Diluted magnetic semiconductor
DMSO	Dimethyle sulphoxide
DD	Double distilled
DRAM	Dynamic random access memory
EDX	Energy dispersive X-ray
EDS	Energy dispersive spectroscopy
EMA	Effective mass approximation
FTIR	Fourier Transform Infra red
FWHM	Full width half maxima
FC	Field cooling
HRTEM	High resolution transmission electron microscopy
HOMO	Highest occupied molecular orbital
IR	Infra red
JCPDS	Joint committee of power diffraction standard
LCAO	Linear combination of atomic orbital
LCAO-MO	Linear combination of atomic orbital-Molecular orbital
LUMO	Lowest unoccupied molecular orbital
LED	Light emitting diode
MFM	Magnetic force microscopy
MO	Molecular orbital

MWNT	multi wall nanotube
MBE	Molecular beam epitaxy
NS	Nanosructure
NC	Nanocrystal
PL	Photoluminescence
PVA	Poly vinyl alcohol
PVP	Poly vinyl pyrrolidone
RTFM	Room temperature ferromagnetism
QD	Quantum dot
SEM	Scanning electron microscopy
SQUID	Super conducting quantum interference device
SPM	Scanning probe microscopy
SRAM	Static random access memory
TEM	Transmission electron microscopy
UV	Ultra violet
XRD	X-ray diffraction
ZFC	Zero field cooling

Introduction

As of today, we notice an ever increasing trend of research in nanoscience and nanotechnology. Nanotechnology [1-5] is considered to be the future miniaturization technology with wide variety of potential applications. Nanotechnology concerns with the study of ultrafine material structures with average crystalline size of the order less than several nanometers. At this quantum-realm scale, quantum mechanical effect is important. Nanostructured materials exhibit unique characteristics due to their existence in the transition region between bulk solid and molecular structure [6-8]. Different nanostructured materials with variable dimensions but having similar chemical composition may behave differently.

In recent years, out of varied nanostructured materials semiconductor nanostructures too have gained considerable interest in research [9-14]. Semiconductor nanocrystals are already commercially utilized as luminescent bio labels [15–17] and have been demonstrated as important components in regenerative solar cells [18–20], optical gain devices [21], and electroluminescent devices [22–24]. Nanometer-scale semiconductor structures have been extensively studied over the last two decades to investigate the influence on physical properties. If the structure is sufficiently small, its boundaries squeeze the carriers and modify their behaviour. In case of semiconductor quantum dot electron and holes are three dimensionally confined inside a nanometer scale "quantum box". The confinement causes a discrete atomic-like energy-level structure for the carriers, and a variety of new physical phenomena are observed [25].

1.1 Quantum confinement effect

Quantum confinement effects in semiconductor nanostructures play an important role to explore many properties which are expected for practical applications. The quantum size effect that dominates in the low dimensional structures with large surface to volume ratio results in band gap enhancement along with discrete energy levels.

Quantum confinement is most commonly associated with nanostructured materials in the sense that bulk materials generally exhibit continuous absorption and electronic spectra. However, upon reaching a physical length scale equivalent to or less than either the exciton Bohr radius or deBroglie wavelength, both the optical and electronic spectra become discrete and more atomic-like. In the extreme case of quantum dots, confinement occurs along all three physical dimensions, x, y, and z such that the optical and electrical spectra become truly atomic-like. This is one reason why quantum dots or nanocrystals are often called artificial atoms [25]. Analogies comparing the particle in a one dimensional box to a quantum well, the particle in a two dimensional box to a quantum wire and the particle in a three dimensional box to a quantum dot provide only half the solution. If one considers that in a quantum well only one dimension is confined and that two others are "free", there are electronic states associated with these extra two degrees of freedom. Likewise in the case of a quantum wire, with two degrees of confinement, there exists one degree of freedom. So solving the particle in a two dimensional box problem models the electronic states along the two confined directions but does not address states associated with this remaining degree of freedom.

In the early 1980s, quantum confinement effect on small particles in suspension was first reported by Ekimov, Efros and Papavassiliou [26,27,28]. Later, proper framework for clear understanding of such effects was laid out by Brus et al. [29,30] on the basis of molecular quantum physics. The motion of electrons, holes and excitons (a quasi-particle, a electron hole pair interacting each other via coulomb potential) are restricted due to finite size of the nanocrystals. This is one of the reason for which nanostructured materials (NSM) exhibit quantum confinement effect.

The excitons correspond to a hydrogen like bound state of an electron-hole pair and characterized by an exciton Bohr radius as defined by [31]

$$a_B = \frac{4\pi\varepsilon_o\varepsilon_r^2}{m_o e^2}\left(\frac{1}{m_e^*} - \frac{1}{m_h^*}\right) \qquad\qquad \rightarrow 1.1$$

where ε_r = relative dielectric constant (high frequency)

m_e^* = effective mass of electron

m_h^* = effective mass of hole

m_o = rest mass of electron

For most of the semiconductors the exciton Bohr radius a_B is about 1-10 nm, it is considerably larger than the respective value for a hydrogen atom (~0.53 Å). The excitonic Bohr radius a_B is an important parameter to explain quantum confinement effect. Quantum confinement effects arise when the size of a nanocrystal is comparable to the length parameters i.e., the deBroglie wavelength λ and exciton Bohr radius a_B of the carriers (electrons, holes, excitons). The movement of electrons, holes and excitons is confined when the radius of a particle approaches the excitonic Bohr radius a_B and coulomb interaction is increased, thereby increasing excitonic binding energy. As a result drastic changes in the electronic structure of nanostructured material are observed. For example one can observe the shift of the energy levels to higher energy, the development of strong oscillator strength between selective transitions and the development of discrete feature of the spectra etc.

When the size of the semiconductor particle approaches Bohr radius of the exciton, the electron hole pair gets spatially confined and assumes a state of higher energy. In this regime of spatial confinement of the charge carriers, the kinetic energy becomes quantized and the energy bands will split into discrete levels. This phenomenon is known as quantum size effect or quantum confinement effect [32]. The particles with reduced dimensions exhibit size-dependent optical and electronic behaviour resulting from three-dimensional (3D) carrier confinement [33]. The blue shift in the optical absorption spectra is observed due to increase in band gap of semiconductor nanostructures.

Quantum size effect can be explained and understand quantitatively with the help of few approaches. The first approach is based on the Effective Mass Approximation [27] (EMA). This EMA model based on:

1. The crystal structure of nanoparticles (Quantum dot) is same as that of the bulk material.
2. The potential barrier at the surface of the Quantum dot if infinite.
3. The Quantum dot is assumed to be spherical.

Quantitatively the quantum confinement effect can be explained with the help of simple particle in a box model.

The solution of the Schrödinger equation gives the Eigen functions

$$\Psi_n(x) = \sqrt{\frac{1}{2L}} \sin(kx) ; \; k_n = \pi n/L \qquad \rightarrow 1.2$$

The corresponding energy Eigen values are given by

$$E_n = \frac{\hbar^2 k_n^2}{2m} = \frac{\hbar^2 \pi^2 n^2}{2mL^2} \qquad \rightarrow 1.3$$

From equation (1.3) it is observed that the energy level spacing is inversely proportional to the square of the length of the box L^2. So, with the reduction of the size of the box, the energy level spacing increases.

According to EMA the band gap energy (E) of a semiconductor nanoparticle can be obtained as

$$E = E_g + \frac{\pi^2 \hbar^2}{2R^2}\left(\frac{1}{m_e^*} + \frac{1}{m_h^*}\right) - \frac{1.8e^2}{\in R} + smaller \; terms \qquad \rightarrow 1.4$$

Where R represents the radius of the semiconductor nanoparticle, E_g is the band gap of the bulk semiconductor. The second and third terms in equation (1.4) represent the quantum localization energy and Coulomb energy with dependence $1/R^2$ and $1/R$ respectively. For large value of R, the energy E, approaches that of E_g. Although the EMA is not valid for very small particles, equation (1.4) has often been used to explain quantum size effects in semiconductor nanocrystallites.

In the second approach the quantum size effect is explained with the help of linear combination of atomic orbital-molecular orbital (LCAO-MO). This approach is widely used to explain the band structure of nanostructures. In this approach nanometer size semiconductor particles are considered as very large molecules. In this method, the overall wave function in a nanoparticle is considered from wave function of individual atomic orbit. This approach is qualitatively illustrated in Figure: 1.1. In case of large crystals the large numbers of atomic orbital's as well as molecular orbital's form energy bands. The highest occupied molecular orbital (HOMO) forms the valance band while the lowest unoccupied molecular orbital (LUMO) forms the bottom the conduction band. The energy difference between the HOMO and LUMO is the band gap of the material. It is observed from the figure: 1.1 that the band gap increases and the bands split into discrete energy levels with a decreasing number of atoms or reduction in size. As the number of atom increase, the discrete energy band structure changes from large energy steps to small energy Contrary to the effective mass approximation the LCAO-MO approach provides a way to calculate the energy structure of very small semiconductor clusters.

As per equation (1.4) it is clear that when the radius of a semiconductor particle decreases its band gap increases. As a result of the spatial confinement of the charge carriers, the kinetic energy becomes quantized. The quantization of the kinetic energy manifests itself as a gradual transition of continuous energy bands to discrete energy levels.

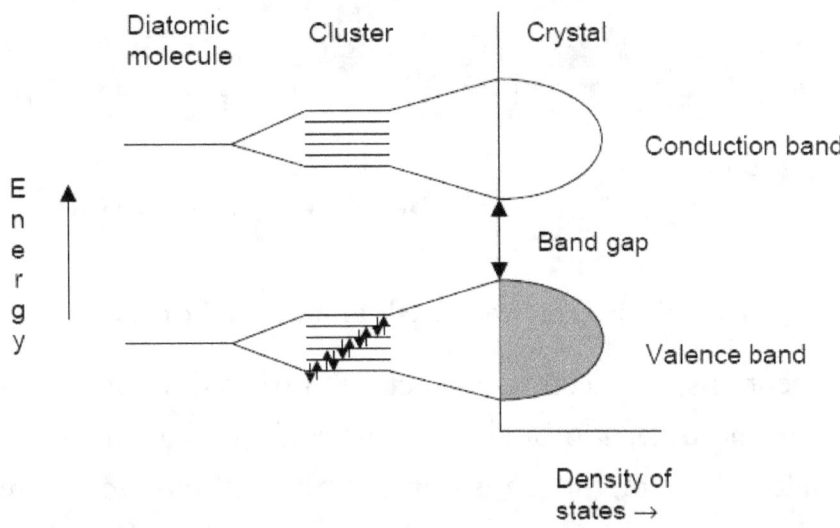

Figure 1.1: Evolution of molecular orbitals into bands according to LCAO-MO

In this tbook, attempt is taken focus on the study of II-VI based transition metal doped diluted magnetic semiconductor nanostructures and their applications in spintronic / Luminescent devices. So, in the next sections we will try to reflect few basic properties of transition metals, study of II-VI semiconductors and few spintronic devices.

1.2 Transition metals

The word *transition* was first used to describe the elements now known as the d-block by the English chemist Charles Bury in 1921, which referred to a transition series of elements during the change of an inner layer of electrons from a stable group of 8 to one of 18, or from 18 to 32. As per IPUAC definition, a transition metal is an element whose atom has an incomplete d sub-shell, or which can give rise to cations with an incomplete d sub-shell. Most of the scientists describe that a "transition metal" is any element in the d-block of the periodic table (all are metals), which includes groups 3 to 12 on the periodic table. In actual practice, the f-block lanthanide and actinide series are also considered transition metals and are called "inner transition metals".

The elements of groups 3–12 are now generally recognized as transition metals, although the elements La-Lu and Ac-Lr and Group 12 attract different definitions from different authors. The positions of the transition metals in the periodic table are presented in figure: 1.2.

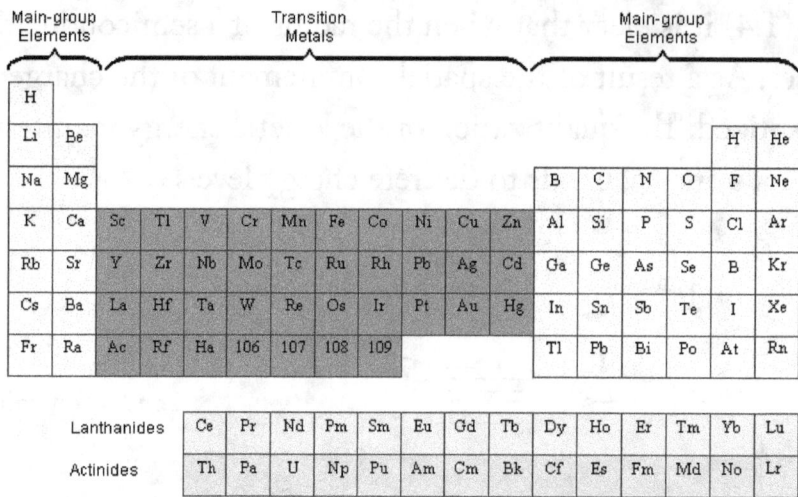

Figure 1.2: Transition metals in the periodic table

With a few minor exceptions, the electronic structure of transition metal atoms can be written as $[\ ]ns^2(n\text{-}1)d^m$, where the inner d orbital has more energy than the valence-shell s orbital. In divalent and trivalent ions of the transition metals, the situation is reversed such that the s electrons have higher energy. Consequently, an ion such as Fe^{2+} has no s electrons: it has the electronic configuration $[Ar]3d^6$ as compared with the configuration of the atom, $[Ar]4s^2 3d^6$.

Transition metals are very hard, with high melting points and boiling points. Moving from left to right across the periodic table, the five d orbitals become more filled. The d electrons are loosely bound, which contributes to the high electrical conductivity and malleability of the transition elements. The transition elements have low ionization energies. They exhibit a wide range of oxidation states or positively charged forms. The positive oxidation states allow transition elements to form many different ionic and partially ionic compounds. The formation of complexes causes the d orbitals to split into two energy sublevels, which enables many of the complexes to absorb specific frequencies of light. Thus, the complexes form characteristic colored solutions and compounds. Complexation reactions sometimes enhance the relatively low solubility of some compounds. these elements do not occur naturally and have not yet been found as the product of a nuclear reaction. Many of the properties of the transition elements are related to the fact that, in their electron structures, the occupied s and d sublevels of highest energy are very close in energy.

1.3 Semiconductor Nanostructured materials

A semiconductor material has conductivity between that of a conductor and an insulator. A typical semiconductor has an electrical conductivity $10^3 Ohm\ cm^{-1}$ to $10^{-8}Ohm\ cm^{-1}$ which lies between that of a conductor and insulator. Semiconductors exhibit a range of useful properties such as unidirectional motion of current. By controlled addition of impurities or by the application of electrical fields or light the conductive properties of semiconductors can be modified, for which they are very useful in device applications. Germanium (Ge) and Silicon (Si) are considered as elemental semiconductors; they exhibit narrow indirect band gap. Binary compound semiconductors are compounds of two different group elements, for example, (III-V): GaAs, NaN, InP etc.; (II-VI): ZnS, CdS, ZnO etc.; (IV-VI): PbS, PbSe, PbTe etc.. It is to be noted that the energy band gap for II-VI based semiconductors are found wider while that for the IV-VI system, it is found to be narrower. Ternary alloy semiconductors are alloys of two binary compound semiconductors, for example, Aluminium gallium arsenide ($Al_xGa_{1-x}As$), Indium gallium arsenide ($In_xGa_{1-x}As$), Aluminium gallium phosphide ($Al_xGa_{1-x}Ps$) etc. Quaternary alloys semiconductors are consisting of four elements, for example, Indium gallium arsenide nitride (InGaAsN), Aluminium gallium arsenide phosphide (AlGaAsP) etc..

In the last two decades, semiconductor nanostructured materials have attracted much more attention owing to their potential in developing devices for future applications. Especially II-VI based semiconductors (e.g. ZnO, ZnS, CdSe, CdS, CdTe, etc.) have attracted much more attention in this regard.

Any semiconductor material containing clusters or grain with dimension below 100 nm may be considered as semiconductor nanostructure. Due to size effect and surface effect nanostructured materials exhibit specific properties. The size effects result from the special confinement of carrier motion in a low dimensional-system. For example the confinement of electron wave functions inside a region whose size is smaller than the electron mean free path may give birth to completely new properties. Most of the atoms in a nanostructured material (NSM) are displayed on the surface than the core, hence surface reactivity increases at large. This surface effect is considered as important origin of specific properties exhibited by NSM. At low dimension few dramatic properties exhibited by NSM are:

a. High surface to volume ratio (>>1) which is ~ 10% for a particle with crystallite size 100Å and 90% for 10Å particle size.

b. Enhancement in band gap energy along with evolution of discrete energy level.

c. Enhancement in the oscillator strength with respect to the bulk.

d. Excitonic binding energy increases with respect to the bulk as a result of which the exciton absorption at the room temperature can be visible.

e. In NSM, most of the atoms are displayed on the surface rather than the core of the particle, surface reactivity increases at large.

Semiconductor nanostructures are considered to be vital components in the next generation technology for their unique size and shape dependant tunable physical properties [25]. It has been reported that doped semiconductor nanocrystals form a new class of materials, which have a wide range of application in devices, lasers, sensors and light emitting diodes (LED) [34-37].

1.4 Diluted magnetic semiconductors (DMS)

There is a considerable interest in having a ferromagnetic material that is compatible with existing semiconductor technology. To this end, dilute magnetic semiconductors (DMSs) [38,39], a new class of materials were developed by incorporating a quantity of transition metal atoms into a traditional semiconductor. In recent years DMS have been investigated to explore materials that could exhibit ferromagnetism at room temperature or well above it. This class of material have promising potential of being used in spintronics, optoelectronic or other devices using both charge and spin degrees of freedom of electrons. Optical properties of this class of material may provide some interesting scope with respect to transforming magnetic information into optical signal.

During last decade, research in diluted magnetic semiconducting systems has gained interest due to possible application in nanomagnetics and spintronics for information storage, transport, and processing. They could, for instance combine manipulation of charges and manipulation of spins in what is called spintronics. Research in DMS nanostructures received a big boost when it was discovered that along with carrier confinement, spin confinement is possible in semiconductors doped with magnetic ions. It has been reported that in a DMS, the strong sp-d interactions between the band carriers and the transition metal ion gives rise to large magneto-optical effects [39]. In a magnetic Quantum dot, this sp-d interaction takes place with a single carrier or a single electron-hole pair. The formation of quasi zero − dimensional magnetic polarons (regions with correlated carrier and magnetic ion spins) has been demonstrated in individual QDs [40]. So far semiconductor quantum dots have been identified as building blocks for spin based solid state quantum logic gates which can function in a fully scalable way [41-43]. The development of such devices require the ability to detect and manipulate individual spins which have been suggested by different groups [44,45]. In addition, incorporation of Mn ions into the crystal matrix of different II-VI semiconductors, successful approaches to fabricate DMS quantum dot and magnet /DMS hybrid structures has been reported [46-49].

Diluted magnetic semiconductor materials are promising component for the new spin based information technology, in which the spin degree of freedom of the electron can be utilized to sense, store, process and transfer information. The past three decades have witnessed the rapid advancement of solid state electronics, including first the replacement of discrete circuit elements and finally the integration of many circuit elements into one semiconductor chip i.e. integrated devices. The only fundamental discrete circuit elements that have been left behind are those involving magnetic materials [50]. Semiconductor and magnetic materials are very important materials in the current electronic industries [51]. Now, the concept of 'magnetoelectronics' i.e. electronic chips consisting of DMS material has come up. One challenge in realizing magnetic RAM involves addressing individual memory elements, flipping their spins up or down to yield the zero and ones of binary computer logic. In order to be practical, magnetoelectronics will need to use semiconductors that maintain their magnetic properties at room temperature. This is a challenge before scientific community as most of the magnetic semiconductors lose their magnetic properties at temperatures well below room temperature, and would require expensive and impractical refrigeration in order to work for actual devices [52].

1.5 II-VI based DMS

II-VI based diluted magnetic semiconductors [53,54] are a class of materials that has received much attention in recent years due to their interesting physical properties and the potential applications in integrated magneto-optical devices [54]. New magneto-electronic devices based on control of the charge as well as the spin of electrons [55] have grown much more interest towards DMS materials. The Curie temperatures are dramatically lowered in the II-VI DMS studied [56,57] compared to III-VI system, but II-VI compounds are ideal materials for fundamental studies. Carrier induced ferromagnetism, as observed in p-type doped Mn based II-VI DMS [58,59], is actually very promising for device applications. One of the remarkable properties of II-VI DMS is the giant Zeeman splitting of the carrier states under application of magnetic field. Moreover, Bulk CdS absorbs in the yellow-green region of the visible spectrum and so used as a window material in solar cells. Reducing the size of CdS down to 40Å, its band gap can be tuned up to UV-region [60]. Similarly, CdSe has a bulk band gap of 1.74 eV which lies in the visible spectrum close to IR region. By decreasing the size of CdSe nanocrystals, one can push the band gap towards the visible region [61]. Also, the band gap of ZnS in the nanometer regime allows us to achieve large quantum efficiency of emission response. In case of impurity (Transition metal and rare earth metal) doped II-VI semiconductors, the impurity centres strongly modify the luminescence properties.

One of the remarkable properties of II-VI DMS is the giant Zeeman splitting of the carrier states under application of magnetic field. This splitting reveals strong exchange interaction between the magnetic ion moment and electronic spins. In a recent work [62] for CdSe/

ZnMnSe self-assembled QD sample the strength of this interaction enhances as a function of magnetic field to become more than 30 meV. Enhancement of Zeeman splitting in excess of the value predicted by standard analysis have been observed in several single quantum well (SQW) structure and superlattices [63-65] whose barriers were made from DMS layers. Recently enhancement of Zeeman splitting in double quantum wells containing ultra thin magnetic semiconductor layers is reported [66]. The enhancement of Zeeman splitting is discussed in terms of effects of reduced dimensionality on the magnetic properties of thin DMS layers. The Zeeman shifts of DQWs in the presence of a magnetic field are calculated using the 8 −band k.p. model, including effect of strain [67-69]. In some II-VI based DMS such as (CdMn)Se, magneto-optic properties were extensively studied, and optical isolators were recently fabricated using their large Faraday effect [70]. Recently, the II-VI compound semiconductors ZnO, GaN have attracted revival attention since it was found that high quality epitaxial thin film display excitonic ultraviolet laser action at room temperature [71-72]. Among all DMS materials Mn doped II-VI[73] and III-V [72] compound semiconductors have been extensively studied. II-VI include a variety of compounds consisting of various combination of group-II cations (Zn, Cd and Hg) and group-VI anions (S, Se and Te), some of which have been applied to magneto-optical devices [74]. III-V compounds can be ferromagnetic materials, where one can control the ferromagnetic properties with electric field or light and can inject spin-polarized carriers into heteroepitaxial semiconductor device [75,76]. However the Curie temperature, T_c has been much lower than room temperature, e.g. $T_{c=}$ 110K in (Ga, Mn)As, hence new DMS having T_c beyond room temperature is desired for future devices. The current high T_c record of 173K achieved in Mn doped GaAs by using low temperature annealing techniques [77-79] is promising, but still too low for practical applications. Recent studies of ferromagnetic Cr and Mn based II-VI DMS were described by T. Dietl [80]. Examples of manipulations with spin ordering by carrier density, dimensionality, light and electric field are given. In recent years, the spin-polarized transport in II-VI DMS heterostructure has been investgated theoretically. Based on a quantum theory and the free electron approximation, Egues [81], Guo et al. [82] and Chang and Peeters [83] studied spin filtering in some II-VI DMS heterostructures. The result showed a strong suppression for one of the spin components of the current density with increasing the external magnetic field. Zhai et al. [84] investigated the effects of conduction band offset on the spin transport in such heterostructures and showed that the positive zero-field band offset can drastically increase the spin polarization. A theoretical study of the dependence of spin-polarized transport on Mn concentration has been reported [85]. Unusual ferromagnetic p-d interactions have been discovered in zinc chalcogenides with paramagnetic ions [86,87]. A theoretical analysis of the kinetic exchange in Cr based $A_{II}B_{VI}$ DMS [88,89] indicated that these interaction should be strongly affected by the static John-Teller effect observed in these materials i.e. the interaction should depend on the relative population of Cr^{2+} ions with different orientations of the tetragonal John-Teller distortions. The antiferromagnetic ion-ion interactions observed in $A_{II}B_{VI}$ DMS impose on upper limit to the increase of the giant free carrier spin–splitting with the growing content of the magnetic ions. The first calculations of the superexchange in Cr-based DMS and the prediction of the ion–ion interactions to be ferromagnetic in these materials have been reported by J. Blinowski & P. Kacman [90]. Within the same model they clarified the origins of the ferromagnetic p-d interactions in Zn chalcogenides with Cr^{2+} paramagnetic ions. Direct magnetization measurement and magnetic circular dichroism owing to short-range superexchange interactions, the spin-spin coupling is merely antiferromagnetic in II-VI DMS. However, a net ferromagnetic superexchange was predicted for Cr-based[90] II-VI DMS. In the initially studied samples [91,92], the Cr concentration was too low to tell the character of spin-spin interactions. A more recent work on (Zn, Cr)Te [93-95] has lead to the observation ferromagnetism by both direct magnetization measurement and magnetic circular dichroism (MCD).

In this research, the properties of transition metal doped II-VI based systems are studied as the material dimensions are reduced to nanometre length scales. A systematic study on the fabrication and properties (structural, magnetic and optical) of a series of ZnO and ZnS samples, which are doped with various transition metal ions (TM= Mn, Co, Cu, Ni, Cr,) has been carried out. Out of these, ZnO:Mn and ZnO:Co systems were extensively studied in this research work.

The main scope of the book was to prepare and investigate samples that could exhibit room temperature ferromagnetism and have good structural and optical properties, which would make them good candidates for practical applications. Since future technology is believed to be dependent on nanoscaled material and devices, it is very important to exploit and optimize interesting properties exhibited by such nanomaterials. Quality of DMS structures was the first target of this research with careful analysis of various physical properties. Selection of host materials and magnetic impurities was the initial step of this research.

1.6 Oxide based DMS

Oxide based semiconductors are attractive semiconductors due to many advantages over non-oxide semiconductors, such as:

1. Having wide band gap suited for application with short wavelength light,
2. Transparent for visible light,

3. Can be doped heavily with n-type carrier,

4. Capability to be grown at low temperature even on plastic substrate,

5. Ecological safety and durability,

6. Low cost, etc. These features serve an important role for various applications.

From the viewpoint of DMS, this feature can be promising for strong ferromagnetic exchange coupling between localized spins due to carrier induced ferromagnetism such as Ruderman-Kittel-Kasuya-Yosida (RKKY) interaction and double exchange interaction when localized spin is introduced in the oxide semiconductor. In addition, large electro negativity of oxygen is expected to produce strong p-d exchange coupling between band carriers and localized spins [96].

There are numerous observations of room temperature ferromagnetism in transition metal doped oxide semiconductors. Yet there is a controversy in regard of the observed ferromagnetism. No farm conclusion was reached in the respect of whether the observed ferromagnetism is due to intrinsic property of the materials or comes from the clustering of magnetic impurity. Moreover, most of the reports on room temperature ferromagnetism were based on thin film investigations. At present, few models based on bound polaron and itinerant carrier induced magnetism are proposed in support of theoretical explanation of magnetism in DMS. So far, no firm conclusion has been drawn by different groups regarding how or why magnetism in DMS material works. However, it is clearly reported that magnetic properties of samples depend on fabrication methods, shape and size of the samples, transport properties etc.

1.7 ZnO based DMS

ZnO is also a direct wide band gap II-VI binary semiconductor with three different forms as hexagonal wurtize, cubic zinc blende, and rarely observed cubic rocksalt. The hexagonal wurtzite structure is most stable and ambient conditions and thus most common in nature. It has a hexagonal lattice belonging to the space group P63mc and is characterized by two interconnecting sublattices of Zn^{2+} and O^{2-} such that each Zn ion is surrounded by oxygen tetrahedral and vice-versa (figure: 1.3). Directions of c and a axes are shown in the figure.

The wurtzite ZnO in its bulk form has a direct band gap of 3.37 eV at room temperature (300^0K) with a relatively large exciton binding energy 60 meV (almost three times larger than that of GaN, the most widely used wide band gap compound).

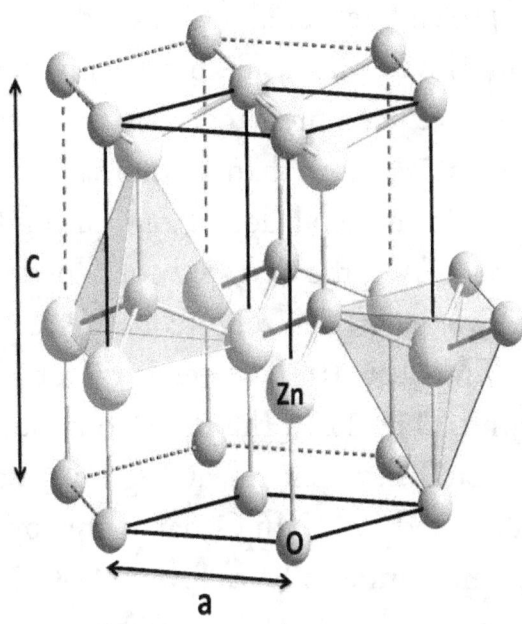

Figure 1.3: Wurtzite structure of ZnO, directions of 'c' and 'a' axes are shown.

Pure ZnO is colorless and clear due to its wide band gap. For significant and specific electrical, optical and acoustic properties of ZnO it has potential applicability for optoelectronic devices such as LEDs, laser diodes and detectors in the UV wavelength range. An important aspect of ZnO is the one dimensional ZnO nanostructures which have attracted special attention owing to their specific properties and diverse hierarchical nanostructures.

Moreover, researchers have found tremendous potential in ZnO regarding its piezoelectric property, biocompatibility and bio-safe nature [97-99] which makes ZnO as an alternate for next generation semiconductor technology. Different techniques such as pulsed laser deposition, sputtering, thermal evaporation, condensation, solid state reaction and chemical methods etc. have been employed to fabricate ZnO based nanostructures.

Many reports have raised serious doubts on the magnetism of magnetically doped ZnO [100] since the results are very sensitive to sample preparation. However, it has been pointed out [100,101] that the lack of ferromagnetism in some samples can be the result of too low density of carriers. In fact in a very recent report, Hong et al. showed that Zn site defects in ZnO thin films give rise to room temperature ferromagnetism, and there is no role for the doped transition metal such as Mn and Fe in introducing magnetism in the ZnO lattice [102]. The systematic variation of magnetism in doped ZnO as a function of the magnetic dopant has been explored in Ref [102], where room temperature ferromagnetism has been found in films doped with Sc, Ti, V, Fe, Co and Ni but not with Cr, Mn, or Cu. Very recently, by analyzing the controlled introduction and removal of Zn, carrier-mediated ferromagnetism in Co-doped

ZnO has been demonstrated [103]. Dietl et al. [104] predicted that Mn-doped ZnO would show ferromagnetic behaviour (FM) with a T_c above room temperature. Several transition metal (TM) doped ZnO films have been prepared since the thermal equilibrium solubility of transition metals in the host metal is higher than 10 mol%. Calculation of electronic structure and magnetism of 3d or 4d transition metal-doped ZnO using full potential linearized plane method [105] has been published. After the first study of Mn-doped ZnO [106], many studies on ZnO doped with various TMs have been reported. Several studies claim nonferromanetic behavior of ZnO doped with TM [106-109], where as the other groups claim ferromagnetic behaviour of the same compounds [110-118]. The reported value of T_c varies from 30K to 550K. For fabrication of Mn doped ZnO different research groups have used different techniques [119-121]. Liu et al. have fabricated $Zn_{1-x}Mn_xO$ nanowires by a chemical vapor deposition method and found the curie temperature to be as low as 44K [122]. However, Philipose et.al. reported room temperature ferromagnetism in (ZnMn)O nanowires grown on silicon substrates fabricated by a vapor-liquid-solid method [123]. The structural and magnetic properties of ZnO:Mn and ZnO:Co nanopowders obtained by soft chemical route were reported [124]. X.Z. Li et. al. reported structural study of Mn-doped ZnO films by TEM in which pulsed laser deposition was used for fabrication [125]. Co doped TiO_2 [121-127] and Mn doped SnO_2 [127] were also reported so far. A few studies report that the precipitation of Co metal is the origin of ferromagnetic signal [114,116]. There have been very few reports on the definite value of T_c because T_c is too high to be measured by conventional tools such as magnetometer employing superconducting quantum interference devices (SQUID). Recent reports establish that magnetically doped TiO_2 [126], ZnO [127] and SnO_2 [128] could represent alternative spintronics materials with T_cs above room temperature and as high as 700K [129]. The anomalous Hall Effect or the hysteretic magnetoresistance seems to be an evidence of ferromagnetism. Ferromagnetic DMS such as $Ga_{1-x}Mn_xAs$ shows anomalous Hall effect. On the other hand, Mn doped ZnO with high carrier concentration shows hysteretic magnetoresistance below 0.2 K implying ferromagnetic ordering, where anomalous Hall Effect is not seen [118].

In addition, it is increasingly evident that the Room temperature ferromagnetism (RTFM) is often observed only in thin film and nanostructures and exhibits notable surface effects [130–132], which is suggested to contact with superficial defects [133,134]. Therefore, nanoparticles (NPs) with the high surface to volume ratio are expected to be an excellent model for studying the defect-related ferromagnetism because most of the defects must exist near the surface of NPs. This implies that the size of NPs should strongly affect the observed ferromagnetism. To my knowledge, however, there is no report on a systematic study on the size dependence of the magnetic properties in pure ZnO NPs. For further progress in this field of research, chief requirements would be:

i.   fabrication of high quality sample,

ii.   in-detail characterizations and

iii.   the elucidation of origin of observed ferromagnetism etc.

1.8 ZnS based DMS

ZnS, an important semiconductor compound of the II-VI group, has wide band gap energy of 3.6 eV at 300K. Zinc sulphide exists with two different crystal structures, the cubic zinc blend (left diagram of figure: 1.3) and the hexagonal wurtzite (right diagram of figure: 1.3). In the zinc blend structure the sulphur ions form an fcc structure and the zinc ions occupy half of the tetrahedral sites in this structure to attain charge neutrality.

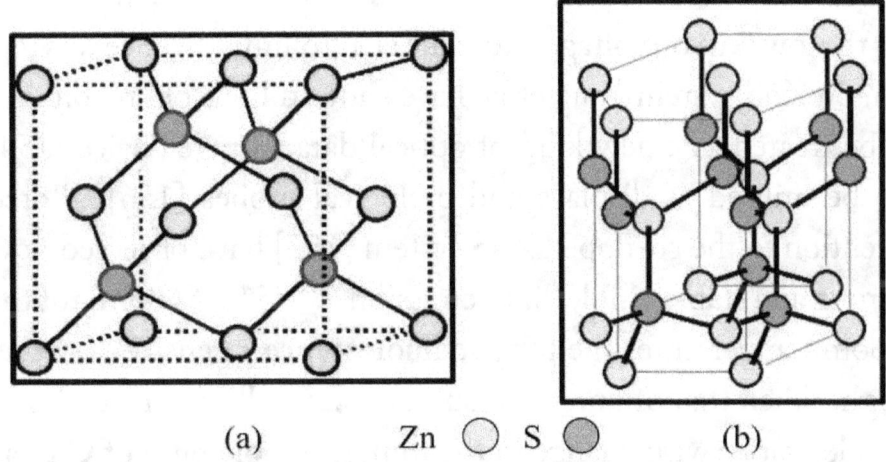

(a)   Zn ◯   S ◯   (b)

Figure 1.4: (a) Cubic Zinc blende and (b) Hexagonal (Wurtzite) structures of ZnS

The ZnS crystal has a lattice parameter of 0.541 nm. The crystal structure is also known as diamond cubic and may be thought of as two interpenetrating fcc lattices, one for sulphur the other for zinc, with their origins displaced by one quarter of a body diagonal (figure: 1.4). The hexagonal wurtzite structure has the sulphur ions in an hcp array and one-half of the tetrahedral sites in this structure are occupied by zinc ions.

The crystal has a basal lattice parameter, a = 0.233 nm and c/a = 1.63 [135]. The phase transition from the cubic form to the wurtize form occurs [136] at around 1020^0. "ZnS" is well known for its wide use in electroluminescence [137] and non linear optical devices. Transition metal (Mn, Cu) doped ZnS semiconductor nanostructures are believed to be potential candidates owing to strong emission due to impurity states [138-140]. Incorporation of both transition metal ions and rare-earth ions ZnS nanostructures by adopting chemical and physical techniques have been reported in recent years [141-143]. Bhargava et al. [144] in 1994, have showed that Mn^{2+} doped ZnS nano semiconductor which are shown to yield both high quantum luminescence efficiency and the lifetime shortening. Since then doped semiconductor nanocrystals have formed a new class of materials resulting a wide range of device applications, light emitting diodes (LED) sensors and lasers [145-148]. Out of numerous works dedicated to different transition metal doped system, Mn^{2+} doped ZnS system has been most extensively studied which have shown orange yellow emission. Mn doped ZnS nanocrystalline films have shown potential applications in thin-film electroluminescence (TFEL) devices. By preparing Mn:ZnS TFEL device, Adachi et al. have reported the existence of reddish-orange broad band emission centred at 626 nm [149]. With Mn:ZnS nanoparticles, Dinsmore et al. [150] have achieved maximum luminescence of ~2.8 cd/m² under low driving voltage. Cu doped ZnS is well known for its property as a good cathode ray tube phosphor which is generally employed for oscilloscopes. This system is also widely studied for possible applications in light emitting diode (LED) [151]. Fabrication of single layer LED using Cu:ZnS nanocrystals/polymer composite having the low turn on voltage with blue electroluminescence was reported by Huang et al. [152]. With Cu:ZnS system Yang et al. have showed high photovoltaic efficiency [153]. Doped ZnS can be treated as a new kind of optical data storage device [154,155]. Mn:ZnS, Eu:ZnS can also be utilised as displays and biological probes [156]. Moreover researchers have also paid attention to the co-doped ZnS system [157] have observed novel luminescence characteristics (strong and stable visible-light emission, λ_{em} ~515-560nm) for the co-doped ZnS nanocrystals at room temperature. The relative fluorescence intensity of the co-doped sample is dramatically higher than that of undoped ZnS nanocrystallites. The emission wavelength of the co-doped samples varies with change in the impurity mole ratio of Cu^{2+} and Co^{2+}.

Due to cheap raw materials, user friendly handling and large scale production [158] chemical precipitation method has been considered as one of the best fabrication technique out of other methods. Different processing parameters like amount of raw materials, reaction temperature, pH value of as prepared solution, titration rate, stirring rate etc. have large effect on different properties of samples prepared. These properties may include particle size, particle size distribution, shape of the particles etc. Bol and Meijerink adopted metallic acetate salts and Na_2S in aqueous solution (pH=10.3) with continuous agitation at room temperature. During fabrication through chemical route the precipitation temperature can be in the range of 0⁰-80⁰ C; pH value within the range 2-10.3; titration rate about 0.5-0.8 ml/s.

Reversed micelles method is a usual preparation technique for ZnS nanoparticles. Xu et al. [159] prepared $ZnS:Cu^{2+}$, $ZnS:Mn^{2+}$, $ZnS:Eu^{3+}$ and Cao et al. [160] prepared core-shell $ZnS:Mn^{2+}/ZnS$ nanoparticles by applying this method of preparation. The reversed micelles are usually thermodynamically stable mixtures of four components: **a)** surfactant, **b)** co-surfactant, **c)** organic solvent and **d)** water. Some usual surfactants are: CTAB (cetyltrimethyle ammonium bromide), SDS (sodium dodeyl sulphate), aerosol-OT (AOT). Precipetation is one procedure to form nanoparticles by the technique of reversed micelles. Two reversed micelles containing the anionic and cationic surfactant are mixed in this method. Since, every reaction takes place in a nanometer-sized water pool, water insoluble naoparticles are formed. Various ZnS nanostructures have been synthesized by the simultaneous in situ formation in liquid-crystal templates, thermal evaporation in the presence of Au catalysts, and through intermittent laser–ablation catalytic growth [161,162], ZnS nanocables and nanotubes were fabricated via a chemical reaction, using ZnO nanobelt as a tamplate and by a thermo chemical process [163,164]; ZnS nanobelts were synthesized by hydrogen-assisted thermal evaporation synbook and also by the thermal evaporation method [165,166].

There is a large possibility to design **spintronic devices** out of transition metal doped II-VI based transition metal doped semiconductor nanostructures as discussed above. The increasing demand of spintronic devices in the electronic technology will certainly change the world's scenario in the next few decades. Already, 'spintroncis' has proved its potential in real applications. In the next sections, applications of few spintronic devecs will be discussed.

1.9 Spintronics

Since possible applications of spintronics is a part of the book we now I look into the specialities of spintronics.

Data storage industry and 'Si' based integrated circuit (ICs) are considered to be the most successful technologies in the world. Since last few decades, development in both the technologies continues to advance at a rapid rate which has changed world's scenario remarkably. A hard disk drive (HDD) is a data storage device used for storing and retrieving digital information using rapidly rotating discs (platters) coated with magnetic material. An HDD retains its data even when powered off. Introduced by IBM in 1956, HDDs became the dominant secondary storage device for general purpose computers by the early 1960s. Continuously improved, HDDs have maintained this position into the modern era of servers and personal computers. As of 2012, the primary competing technology for secondary storage is flash memory in the form of solid-state drives (SSDs). HDDs are expected to remain the dominant medium for secondary storage due to predicted continuing advantages in recording capacity and price per unit of storage; but SSDs are replacing HDDs where speed, power consumption and durability are more important considerations than price and capacity. On the other hands the rapid advancement of ICs technology is obeying Moore law that predicts that the number of the transistor of a chip double every 18 months. All ICs operate on controlling the flow of the carriers of charge (electron and hole) through the semiconductor when an electric field is applied. This is the dominant parameter in this type of devices. For the case of magnetic data storage, the dominant parameter is the spin, which is the fundamental origin of the magnetic moment. ICs are important for high speed signal processing and excellent trustworthy, but the memory elements are volatile (the stored information is lost when the power is switched-off). A big advantage of magnetic memory technologies is that these are non-volatile since they employ ferromagnetic materials which by nature have remanence.

Electron spin can be detected as a magnetic field having one of two orientations, known as *down* and *up*. This provides an additional two binary states to the conventional low and high logic values, which are represented by simple currents. With the addition of the spin state to the mix, a bit can have four possible states, which might be called *down-low*, *down-high*, *up-low*, and *up-high*. These four states represent quantum bits (qubits). The existence of four, rather than two, defined states for a logic bit translates into higher data transfer speed, greater processing power, increased memory density, and increased storage capacity, provided the properties of electron spin can be sufficiently controlled for practical applications.

A new field of the electronic technology opens the possibility for the study and understands of the properties a new material that tries combining these two promise characteristic (charge and spin). This branch of the electronic is known as '**spintronics**' (*spin transport electronics*). Spintronics, or spin electronics, refers to the study of the role played by electron (and more generally nuclear) spin in solid state physics, and possible devices that specifically exploit spin properties instead of or in addition to charge degrees of freedom. Spintronics is an emerging field of nanoscale electronics involving the detection and manipulation of electron spin. Spintronic technology has already been tested in mass-storage components such as hard drives. The technology also holds promise for digital electronics in general. Mission has already started to develope spintronic devices that could be much smaller, (less than 100 nanometers) with consume less electricity and be more powerful for certain types of computations than is possible with electron-charge based system.

Traditional electronic devices use only carriers to transport and process electric signal. The development of electrical industry and technology demands devices with higher performances which should include smaller, faster, less energy consuming devices. The smaller dimension is one of the basic requirements. Spintronic uses both charge and spin degrees of freedom which gives choices and functionalities in electronic devices. Using both spin and charge of carriers one can efficiently shrink the size of devices, lower device power consumption and enhance running speed. In addition to this, one can integrate computation and storage components together, which can be used in high density data storage application. Ferromagnatism in Mn–doped GaAs, the prototypical dilute magnetic semiconductor (DMS), has so far been attributed to hole mediated RKKY- type interactions. Unusual directional dependence of exchange energies in GaAs diluted with Mn has been studied theoretically [167]. Also, a theory has been reported for the spin-contribution to the electron –paramagnetic resonance shift (P_s) for an electronic system in the presence of a periodic potential, spin–orbit (SO) interaction, conduction electron–local moment interaction, and an applied magnetic field [168]. In context of research in DMS, most of the researchers are engaged in theory or modeling / simulation to understand basic insights of physics and feasibility of such devices. However actualization of spin based device depends on the figure of merit (FOM) of the devices based on experiments. Only few groups in our country are involved in experimental works in this emerging area.

Doping is a well-established and a very important technique to tailor properties of bulk semiconductors [169]; doping semiconductor nanocrystals provides the possibility of combining tunable properties of finite sized systems with those of doped semiconductors [170]. The growth mechanism of such nanocrystals in order to carry out synthesis of high quality samples has been investigated [171] and novel ways of understanding their internal structure has been established [172]. Magnetic and luminescent properties of several such II-VI semiconducting nanocrystals by doping them with various transition metal ions have also been investigated [173-175]. ZnO, TiO_2 doped with 3-d transition metal Co, Ni, Fe etc are promising DMS materials showing ferromagnetism with Tc above room temperature [176]. The large band gap may open up transparent ferromagnetic materials in oxide DMS for application in the visible region. Band gap measurement of ZnO doped with transition metal Mn, Co and Ni and the variation of band gap with dopant concentration has been explained recently [176]. It is now apparent that without the introduction of Mn or Co; ZnO cannot exhibit ferromagnetism [177]. Synthesis of ZnO doped with Cr by co-precipitation and ceramic methods has been reported [178]. The research group used a vibrating sample magnetometer in which an indication of ferromagnetic-like behaviour at 300K and a spin glass state at 77k have been reported and the research group concluded that the co-precipitation method is more convenient for obtaining single-phase compound by the relatively low temperature processing of the precipitated hydroxide. Very recently, synthesis and magnetic properties of Mn doped ZnO nanowires have been reported by another research group [179]. The group prepared DMS by using a simple auotocombustion method and observed that an increase in the hexagonal lattice parameters of ZnO on increasing the Mn concentration.

Synthesis and proper doping method is a vital part for characterization of samples. Different research groups may have used different techniques for fabrication but inexpensive chemical rout [180-182] draws basic interest towards fabrication of DMS nanostructure materials.

Compatible for nanoelectronics technology it is highly desirable to turn a semiconductor into a ferromagnet. The essential basic conditions required for this are:

a. Starting with a semiconductor which can be doped, and which can be grown epitaxial with good interfaces, thus allowing to control local electric fields in order to manipulate the free carriers in the conduction or valence band;

b. Introducing magnetic impurities (Transition metal) which carry a local magnetic moment to an incompletely filled d-shell;

c. Having a strong interaction between the spin of the magnetic impurities and the free carriers.

Besides these three conditions, other conditions must be added, depending on different goals and objectives. Fulfilling these three conditions allows us to add new magnetic properties to the specific properties of semiconductors. It results in strong magneto-transport and magneto-optical properties, which can be turned into functions of interest to spin-electronics and spin-photonics, but also to original functions such as the control of the magnetic properties by an electric field.

1.9.1 Spintronics Devices

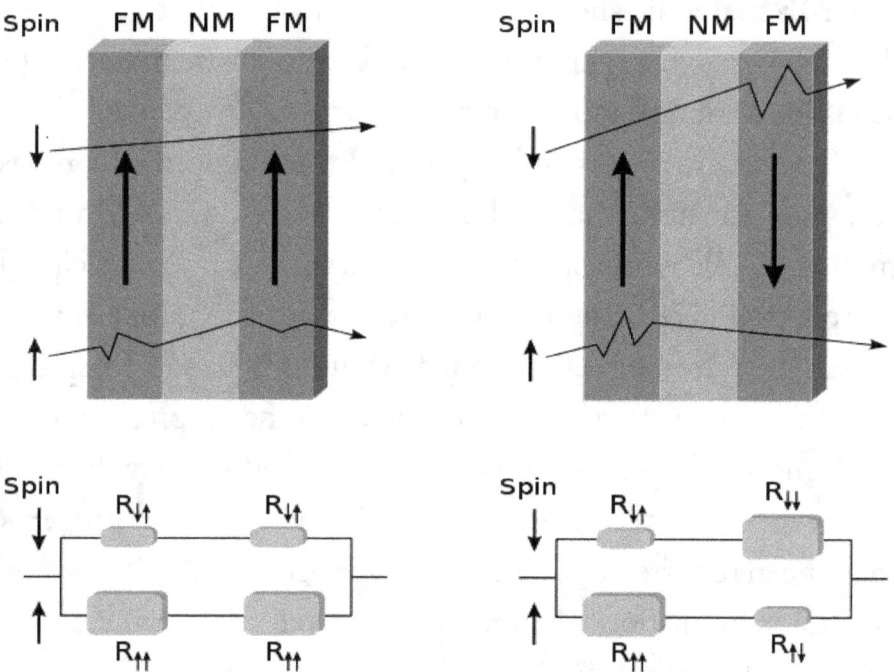

Figure 1.5: GMR effect

The prototype device that is already in use in industry as a read head and a memory-storage cell is the giant-magnetoresistance 'GMR' (183-185) sandwich structure which consists of alternating ferromagnetic and nonmagnetic metal layers. Depending on the relative orientation of the magnetizations in the magnetic layers, the device resistance changes from small (parallel magnetizations) to large (antiparallel magnetizations). This change in resistance (also called magnetoresistance) is used to sense changes in magnetic fields. The GMR effect can be well understood from the figure: 1.5, where two two ferromagnetic layers are kept parallel and antiparallel to get the device resistance low and high accordingly.

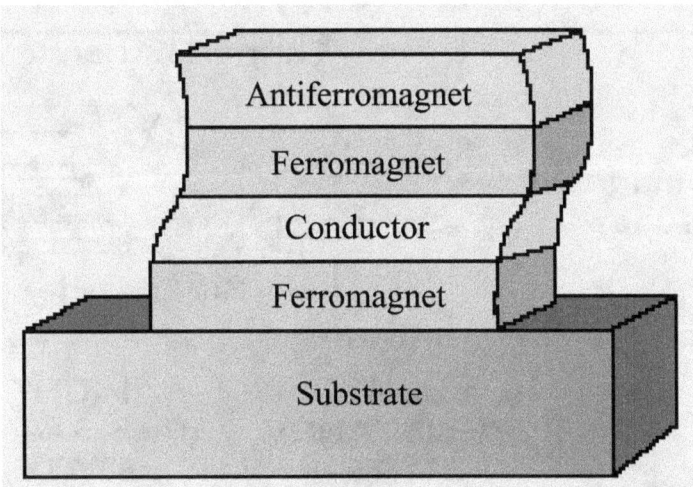

Figure 1.6: Spin valve structure based on GMR effect [Ref:20]

The Schematic representation of first significant spin valve based GMR effect [186] is shown in Figure: 1.6. This consist in tri-layered sandwich structure, where a 'soft' ferromagnetic layer responds to a magnetic field, while a pinned ferromagnetic layer does not. A thin conductor layer (this layer is nonmagnetic normally cooper) is placed between the two ferromagnetic layers. When the magnetizations in the two ferromagnetic layers are parallel, conduction electron pass between them more freely than when the magnetizations are anti-parallel, and thus the resistance is lower in the parallel magnetization case than in the anti-parallel case. This change in resistance (also called magnetoresistance) is used to sense changes in magnetic fields. Almost ever these devices are accomplished through coupling to an antiferromagnetic layer (typically constructed from iron and manganese), as shown in the figure: 1.6. This structure can be configured as a magnetic field sensor, for example in automobile braking system, hysteretic memory device (non-volatile memory chips) or as a magnetic tunnel junction device. The GMR heads are more superior and powerful to conventional MR heads for the reason that they are more super-sensitive, can detect weaker and smaller signals, which does to increase *a* real density (also called bit density) and thus capacity and performance.

In a spintronic device, GMR sensor structures consist of two ferromagnetic alloys sandwiched around an ultrathin nonmagnetic conducting middle layer. The first magnetic layer polarizes the electron spins, while the second layer scatters the spins strongly if its moment is not aligned with the polarizer's moment. If the second layer's moment is aligned, it allows the spins to pass.

The resistance therefore changes depending on whether the moments of the magnetic layers are parallel (low resistance) or antiparallel (high resistance).

(a) Magnetic Tunnel Junction and MRAM (MTJ)

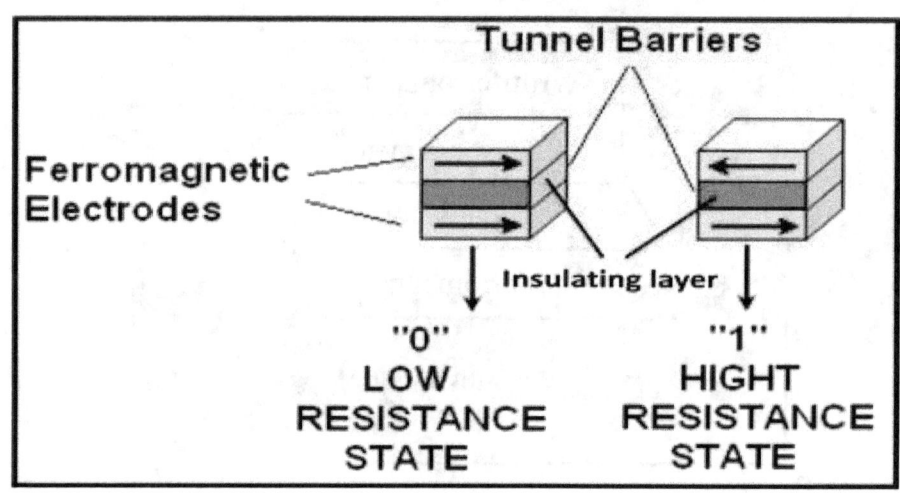

Figure 1.7: Schematic representation of a MRAM with MTJ.

After spin valve the magnetic tunnel junction (MTJ) is the second type of spintronics device. The basic MRAM cell is the so-called Magnetic Tunnel Junction (MTJ) structure of which consists of two magnetic layers sandwiching a thin (sub-nm) insulating layer (Figure: 1.7). Commonly aluminium oxide is used as insulating layer. The electrons can tunnel through the insulating layer and additionally the probability, of tunnelling from a ferromagnetic electrode depends of the spin direction, the resistance of the MTJ is different for the parallel and antiparallel orientation of the magnetic moments of the electrodes. MTJ are very small size, below the micron range and can be fabricated by lithographic techniques. An important application of this small size MTJ will be for a new type of computer memory.

As presented in the Figure 3, the magnetization of one of the layers, acting as a reference layer, is fixed and kept rigid in one given direction. The other layer, acting as the storage layer, can be switched under an applied magnetic field from parallel to antiparallel to the reference layer, therein inducing a change in the cell resistance. The corresponding logic state ("0" or "1") of the memory is hence defined by its resistance state (low or high), monitored by a small read current. A fully functional MRAM memory is based on a 2D array of individual cells, which can be addressed individually. These MRAM is presently in development and it is expected to reach similar densities and access times as the current DRAM or SRAM, but the their main objective on these volatile semiconductor-base memories is that they can retain data after the power is turned off, possibly eliminating the long boot-up time when the computer is switched on. This technology has been used efficiently in the first generation of MRAM devices, developed so far with feature sizes greater than 0.13-0.18 µm [187].

(b) Spin FET:

The idea of spin based field effect transistor (spin FET) was first proposed by Datta and Das in 1990 [188]. The scheme is illustrated in figures: 1.8 and 1.9. Like usual FET, the structure consists of a drain, a source, a narrow channel, and a gate for controlling the current. The gate allows the current to flow (ON) or does not (OFF). In the Datta-Das Spin FET the source and the drain are ferromagnets acting as injector and detector of the electron spin, the drain injects electrons with spins parallel to the transport direction. The electrons are transported ballistically through the channel. When they arrive at the drain, their spin is detected.

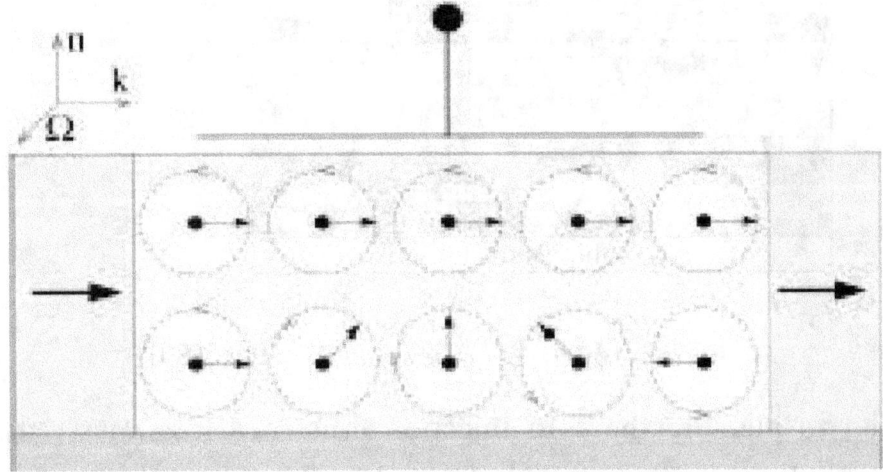

Figure 1.8: Scheme of the Datta-Das spin field-effect transistor (SFET) [from ref: 188].

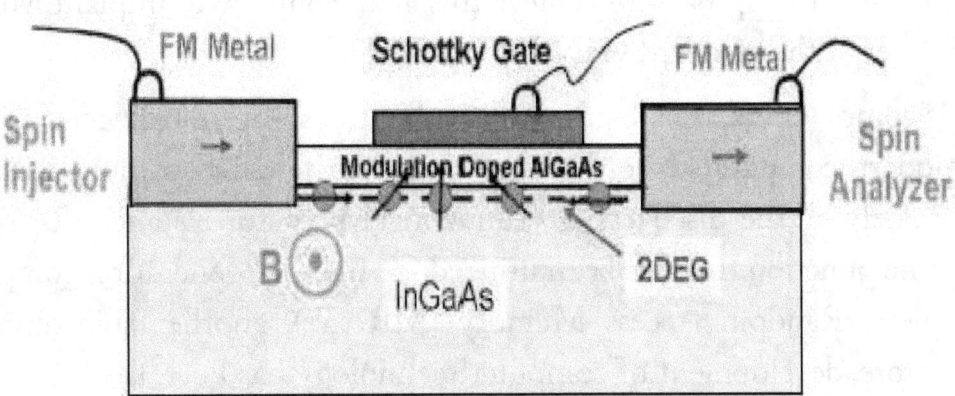

Figure 1.9: Scheme of the Datta-Das spin field-effect transistor (SFET) [from ref: 188].

The electron can enter the drain (ON) if its pin points in the same direction as of the spin in the drain, if not it is scattered away (OFF). The role of the gate is generating an effective magnetic field in the direction as shown in figure. This effective magnetic field causes the electron spins to process. By modifying the voltage, one can cause the precession to lead to either parallel or antiparallel electron spin at the drain, effectively controlling the current.

(c) Spin LED

Efficient spin injection from the ferromagnetic metal and dilute magnetic semiconductor into semiconductor is the fundamental requirement of the semiconductor-based spintronics device.

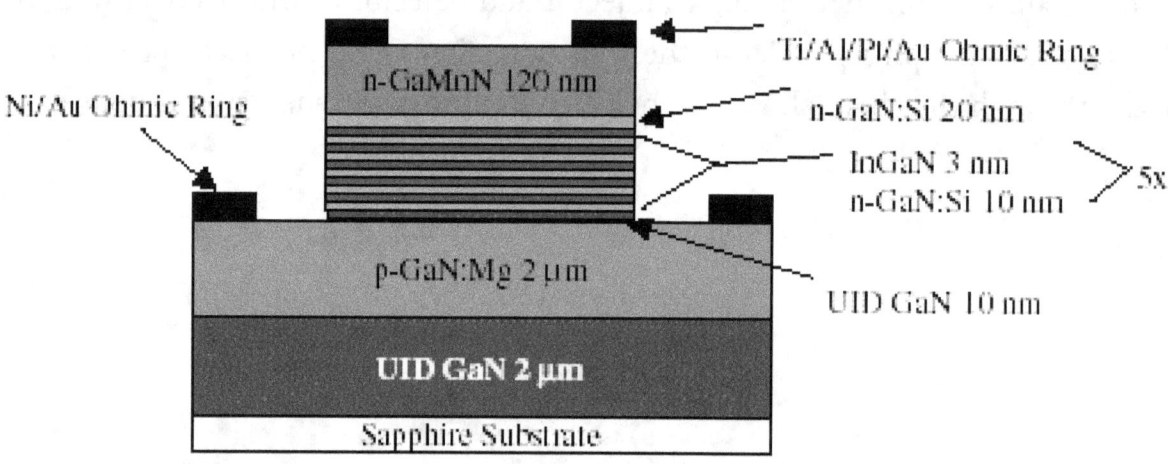

Figure 1.10: A schematic spin LED [Ref: 189]

Spin-LED is used to measure the spin injection efficiency into materials. In a spin Led circularly polarized light is emitted after the recombination of spin polarized carriers that are electrically injected into semiconductor structure. The polarization of the light emitted by the spin LED can be modulated by application of an external magnetic field (Figure: 1.10). The most straightforward approach to achieve this goal would be to implant Mn into the top contact p-GaN layer of the standard GaN/InGaN LED.

Today, at this computer era, we are practically using the spintronic devices; since, spin valve is used in all modern computers in order to read and write data on their hard drive. It was followed immediately by the discovery of Tunneling Magnetoresistance (TMR) leading to the magnetic tunnel junction that has been utilized for the next generation computer memory known as Magnetic Random Access Memory (MRAM), another spintronics device for computers. Therefore, development of computer technology has been the initial driving force for spintronics research.

1.10 Objectives of the present study

Since future technology is believed to be dependent on nanoscaled material and devices, it is very important to exploit and optimize interesting properties exhibited by such nanomaterials. Further, spin based nanostructures are important for deployment of transfer of quantum information from photon system to the spin of the electron in the semiconductor and therefore, detail exploration with regard to power consumption, operational speed and stability of the device can be assessed by the quality of the DMS material. Hence, quality of DMS structures was the first target of this research with careful analysis of various physical properties. Selection of host materials and magnetic impurities was the initial step of this research. Fabrication of DMS was done by using inexpensive chemical method [182,190]. Preliminary studies by using X-ray diffraction, Electron microscope were used for crystallite size estimation etc. Luminescent spectroscopy was studied to detect additional emission line due to incorporation of magnetic ion(s). M~ H loop (hysteresis) was carried out to compare various magnetic parameters. Low temperature M~T measurement were carried out to study the magnetic nature of DMS samples. Atomic force microscopy and Magnetic force microscopy was attempted to visualize magnetic domain.

In this book a systematic study on the fabrication and investigation of structural, magnetic and optical properties of several series of ZnO and ZnS samples, which are doped with various transition metal ions (TM= Mn, Co, Ni, Cr). Out of these, transition metal (Mn, Co) doped 'ZnO' DMSs were extensively discussed. The main scope of this book is preparation and investigation of samples with good structural and optical properties that could exhibit room temperature ferromagnetism, which would make them good candidates for practical applications.

The overall objectives of the book can be summarized as follows:

i.   *To fabricate Transition metal doped II–VI based diluted magnetic semiconductor (Mn, Co, Ni doped ZnO and Mn, Cr doped ZnS) nanostructurs by adopting low cost chemical fabrication route.*

ii. *To exploit and optimize interesting properties exhibited by such nanomaterials.*

iii. *To analyse structural, optical and magnetic properties of as-fabricated DMS materials.*

iv. *To fabricate high quality DMS materials, that can support future spin based or optoelectronic devices.*

v.   *To discuss possible applications in spintronic / luminescent devices.*

Fabrication of DMS Nanostructures

Fabrication of DMS nanostructures as described in this chapter was carried out by adopting inexpensive chemical synthesis method. We present here the result of my current efforts on fabrication of transition metal (TM) doped ZnS and ZnO nanocrystals through chemical as well as solid state chemical reaction route respectively. Synthesis of cetyl trimethyle ammonium bromide (CTAB) assisted TM (Mn, Co, Ni) doped ZnO nanostructures were carried out by adopting solid state chemical reaction route at room temperature. We found elongated nanostructures for all TM:ZnO systems in the form of powder. Study was carried out by fabricating TM:ZnO samples with three different TM concentrations ranging from 1% to 5%.

Gamelin *et al* [191] reported the synthesis of colloidal Mn^{+2} doped ZnO (Mn^{+2}:ZnO) quantum dots at room-temperature and the preparation of ferromagnetic nanocrystalline thin films by hydrolysis and condensation reaction in DMSO under atmospheric conditions. Gamelin's group proposed that direct chemical syntheses of ZnO DMSs can provide better control over material composition than is obtained with some high-temperature vacuum deposition or solid-state synthesis techniques. The same group reported robust, high-Tc in thin films of these nanocrystals prepared by spin coating at 300 K. The corresponding coercivity was 92 Oe, approximately.

Polyvinyl alcohol (PVA) embedded Mn:ZnS and Cr:ZnS spherical nanostructures have been developed at room temperature by adopting low cost chemical synthesis route. As synthesised TM:ZnS nanostructures were found in the form of colloidal solution. Study of Mn:ZnS system was done by varying Mn concentration ranging from 0.008% to 0.25%.

Along with PVA, another surfactant Polyvenyl Pyralidone (PVP) has been utilised to develop Cr:ZnS system.

Only few synthesis processes of colloidal nanoparticles have been reported in the literature. Colloidal nanocrystals should form ideal hosts for strongly-confined artificial atoms, i.e. configuration of one to a few electrons occupying the conduction energy levels in the nanocrystal. Such nanocrystals can be obtained in molar quantities by wet chemical synthesis with increasing control of the size, shape and surface electronic properties.

2.1 Materials

From the application point of view, the large scale synthesis and long term preservation of as synthesised nanoparticles are very much important. In order to increase the stability of the nanoparticles and to protect from the environmental attack they are to be embedded with glass, zeolite or polymer [192]. Surfactants usually play an important role in the electrolysis process to bring some properties of the electro-deposited layer, such as the brightness, moistness, smoothness, and homogeneity.

(i) Cetyl trimethyle ammonium Bromide (CTAB)

For fabrication of Transition metal doped ZnO nanostructures we have utilized CTAB as a surfactant. The surfactant CTAB (cetyl trimethyl ammonium bromide) is a cationic surfactant. It was considered as having good potential in the application of nano-technology because of its strong adsorbed ability in nano particles [193-195]. Besides, the CTAB is also suitable for being as a corrosion inhibitor [196-198]. The two ends of each CTAB molecule has two specific properties, one end with high positive property is hydrophilic (head), while the other end is hydrophobic (tail). In water the hydrophilic end (ion head) repeal each other due to long range electrical force and hydrophobic end (tail chain) attract each other due to short range Vander-wall force promoting micelle formation by the CTAB molecule. The number of molecules present in a micelle once, the critical micelle concentration (CMC) has been reached. At room temperature (~30°C), CTAB forms micelles with aggregation number 75-120.

(ii) Polyvinyle Alcohol (PVA/PVOH)

Polyvinyl alcohol (PVA) is an odourless and tasteless, translucent, white or cream colour granular powder. Generally, it is used as a moisture barrier film for food supplement tablets and for foods that contain inclusions or dry food with inclusions that need to be protected from moisture uptake.

By adopting chemical synthesis route we have fabricated polyvinyl alcohol (PVA) assisted Mn and Cr doped ZnS nanostructures. PVA has excellent emulsifying, film forming, and adhesive properties. It is also resistant to oil, grease and solvent. It is odourless and nonotoxic. It has high flexibility and tensile strength as well as high oxygen and aroma barrier properties. However these properties are dependent on humidity. PVA absorbs more water with higher humidity. The water, which acts as a plasticiser, will then reduce its tensile strength. PVA is fully degradable and a quick dissolver. However, in cold water, it dissolves slowly, but with application of heat it dissolves quickly. PVA has a melting point of 230^0C and $180\text{-}190^0C$ for the fully hydrolysed and partially hydrolysed grades. It decomposes rapidly above 200^0C as it can undergo pyrolysis at high temperatures. PVA is an atactic material but exhibits crystallinity as the hydroxyl groups are small enough to fit into the lattice without disrupting it.

PVA is a good protective colloid for aqueous emulsions and is employed for this purpose in a large variety of emulsion and suspension systems. It also finds use in wet strength adhesives.

The PVA matrix is presented as,

(iii) Polyvinyle Pyrrolidone (PVP)

We have also fabricated Cr doped ZnS nanostructures by utilizing another polymer matrix Polyvinyle Pyrrolidone (PVP) for further investigation. Polyvinyl pirrolidone (PVP) is an excellent polymer with remarkable combination of properties (details are given in the appendix). It has variety of properties, such as: Transparency, adhesive and cohesive, protective colloid and suspending agent, film former, chemical and biological inertness, low toxicity, high media compatibility, cross linkable flexibility etc..

The PVP matrix is given by:

2.2 Fabrication of Bare ZnO nanostructures

For fabrication of bare ZnO nanostructures the the materials utilized are;

i.   Zinc acetate dehydrate (ZAcD): [$Zn(CH_3COO_2.2H_2O$];

ii.   Cetyle trimethyle ammonium bromide (CTAB): $[C_{19}H_{42}BrN]$;

iii. Sodium hydroxide flakes: NaOH.

All reagents were of analytical grade (99.99 % purity) and were used without further purification. Bare ZnO nanoscale systems were fabricated by adopting a solid state chemical reaction route. Zinc acetate dehydrate (ZAcD), Cetyle trimethyle ammonium bromide (CTAB) and Sodium hydroxide flakes (NaOH) are mixed by keeping their molar ratio as 1:0.5:2. The mixture was ground gently in a mortar at room temperature for few hours (~2 hrs) till a paste like compound is obtained. The mixture was repeatedly washed with double distilled water and then annealed at 60^0-80^0C for about 4-5 hrs. After washing with double distilled (DD) water the as-synthesised product was dried and the final product was obtained in the form of powder.

2.3 Fabrication of TM doped ZnO nanostructures

We have adopted solid state chemical reaction chemical reaction route to fabricate TM(Mn. Co, Cr, Ni) doped ZnO nanostructures. Acetates of Zn and TM taken in the appropriate molar ratio to obtain three samples from each TM as 1at.% TM, 3at.% TM and 5at.% TM. The acetates of Zn and TM were ground gently in a mortar at room temperature for ~ 2 h. Sodium hydroxide flakes and cetyl trimethyl ammonium bromide (CTAB) were mixed in the precursor and grinded for several hours till a paste like compound is obtained. The mixture was repeatedly washed with double distilled water and then annealed at 60^0-80^0C resulting in TM doped ZnO nanostructures in the form of powder. According to the TM concentrations in the samples (1%, 3% and 5%), they were indexed as TM-1, TM-3 and TM-5. A part of the as-prepared powdered sample was kept for XRD, PL, FTIR, EDX and HRTEM studies and another part was kept for pellet preparation to carry out magnetic characterization. The powdered samples were pressed into pellets of about 3-4 g by weight and 10 mm in diameter at a pressure of 6 tons/cm ^{2}.

2.3.1 Fabrication of TM doped ZnS nanostructures

(i) Mn doped ZnS

ZnS:Mn nanoaggregates are fabricated by adopting simple, low cost chemical route using polyvenyl alcohol as desired matrix. For this, 2.5wt% Poly vinyl alcohol (PVA) was prepared by using Double distilled water and Magnetic stirrer with stirring at ~200 rpm at a constant temperature of 65^0 until a trnsparent solution is formed. Zinc chloride(aq) and manganese chloride(aq) solutions were prepared separetely and then they were stirred in a mixed environment with variable Mn: Zn ratio (Mn conc. 5% -.008%).

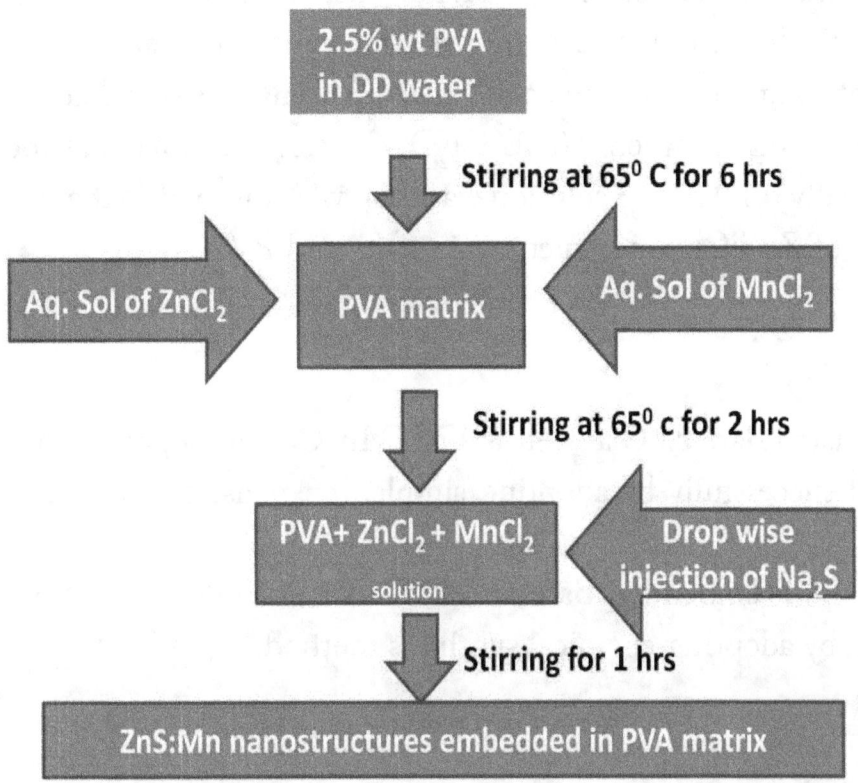

Figure 2.1: Flow chart showing synthesis route of ZnS:Mn nanostructures embedded in Polyvinyl alcohol (PVA) matrix.

The as prepared precursor was mixed with PVA matrix under stirring. Into this dropwise injection of Na_2S (.01M) solution led to the grouth of 'ZnS:Mn' nanoparticles. Figure: 2.1 shows the flow chart of the chemical route for synthesis of the " ZnS:Mn in PVA" system

Out of eight samples prepared with different Mn:Zn molar concentrations within the range 5at.%Mn to 0.008at.%Mn, three samples 0.25at.%Mn, 0.188at.%Mn and 0.008at%Mn were selected on the basis of their prelemenary investigations carried out with UV-Vis ans Photoluminescence spectroscopy and indexed as A, B and C for further investigations. Fluid samples were collected for UV-VIS study and TEM measurements while thin films were casted on glass slides for photoluminescence study, AFM, MFM and XRD investigations.

(ii) Cr doped ZnS

Cr doped ZnS encapsulated in polyvinyl alcohol (-C_2H_4O)n and Polyvinyl pyrrolidone k30 (C_6H_9NO)x was fabricated using a low cost colloidal solution casting route. For fabrication of Cr doped ZnS nanostructures in PVA matrix, we have followed exactly the same route as described above for Mn doped ZnS system, only manganese chloride(aq) has been replaced by chromium chloride(aq). In case of Cr doped ZnS in Polyvinyl pyrrolidone (PVP) matrix, the organic host PVP was taken in place of PVA. For this, 2% (w/v) PVA/PVP) in double distilled water were magnetically stirred at ~200 rpm at a constant temperature for six hours separately until a transparent solution were formed. Next, 0.1 M $ZnCl_2$ was added to the dielectric hosts under stirring condition and then $0.01M$ Cr_2O_3 solution was mixed at room temperature in both matrices. Finally 0.1 M Na_2S solution was drop wise injected to the two precursors, which led to the growth of ZnS:Cr nanostructures in PVA and PVP.

Conclusions

- CTAB assisted Bare ZnO as well as TM (Mn, Co, Ni) doped nanostructures have been fabricated successfully by adopting simple, inexpensive solid state chemical reaction route.
- PVA and PVP embedded Bare and TM (Mn, Cr) doped ZnS nanostructures were developed by adopting chemical synthesis method.

Experimental Techniques

Besides chemical methods, a large variety of experimental techniques were employed to characterize the materials. **Chapter -3** deals in different characterization techniques such as X-ray diffraction (XRD), transmission electron microscopy (TEM), energy dispersive X-ray spectroscopy (EDXS), selected area electron diffraction (SAED), photoluminescence spectroscopy (PL), atomic force microscopy (AFM), magnetic force microscopy (MFM), UV-Visible optical absorption spectroscopy (OAS), Super conducting quantum interference device (SQUID) etc. The actual incorporation of TM atoms into II-VI host structure was also evidenced by the X-ray Diffraction analyses. Electron microscopy reveals formation of spherical nanostructures in ZnS based systems and nanorods in ZnO based systems. Low temperature magnetic measurements of the investigated samples were performed by utilizing SQUID within temperature limit 4K to 300K and at different magnetic fields 0.01T and 0.1T. The characterizations of samples were carried out by utilizing modern experimental tools at different laboratories / facility centers as given below:

Laboratories/ Facility Centers	Tools utilized
1. (a) Tezpur University (TU)	EDX, SEM ;
(a) Dept. of Physics, TU	XRD, UV-Vis, PL;
(b) Dept of Chem. TU	FTIR, Pallet preparation;
2. SAIF, NEHU , Shillong	TEM/HRTEM, SAED;
3. DAE UGC Consortium, Indore	Low temp. SQUID;
4. CMMIP, University "Politehnica" Bucharest, Romania	AFM, MFM.

3.1 X-ray Diffraction

"……every crystalline substance gives a pattern; the same substance always gives the same pattern; and in a mixture of substances each produces its pattern independently of the others."

– By A. W. Hull (1999)

Solid matter can be classified as amorphous and crystalline. In amorphous solid atoms are arranged in a random way similar to the disorder as observed in a liquid. Glasses are amorphous materials. On the other hand, in a crystalline solid the atoms are arranged in a regular pattern. About 95% of all solid materials can be described as crystalline. When X-rays interact with a crystalline substance (Phase), one gets a diffraction pattern. The X-ray diffraction pattern of a pure substance is, therefore, like a fingerprint of the substance. The powder diffraction method is thus ideally suited for characterization and identification of polycrystalline phases. Today, diffraction patterns have been collected and stored on magnetic or optical media as standards for about 50,000 inorganic and 25,000 organic single components, crystalline phases. By following search/ match procedure the powder diffraction is used to identify components in a sample. Furthermore, the areas under the peak are related to the amount of each phase present in the sample.

An electron in an alternating electromagnetic field will oscillate with the same frequency as the field. When an X-ray beam hits an atom, the electrons around the atom start to oscillate with the same frequency as the incoming beam. In almost all directions destructive interference will be obtained, that is, the combining waves are out of phase and there is no resultant energy leaving the solid sample. However the atoms in a crystal are arranged in a regular pattern, and in a very few directions we will have constructive interference. The waves will be in phase and there will be well defined X-ray beams leaving the sample at various directions. Hence, a diffracted beam may be described as a beam composed of a large number of scattered rays mutually reinforcing one another.

X-ray reflections are obtained for different parallel planes (h k l) inside a crystal according to their orientation. When the value of any indices becomes zero, the plane is parallel to that axis. For e.g., (2 2 0), the plane is parallel to c axis.

The X-ray Diffraction can be easily understood with the help of Bragg's law [199]. The constructive interference exhibited by the diffracted X-ray from two parallel planes with interplanar spacing d is shown in figure: 3.1

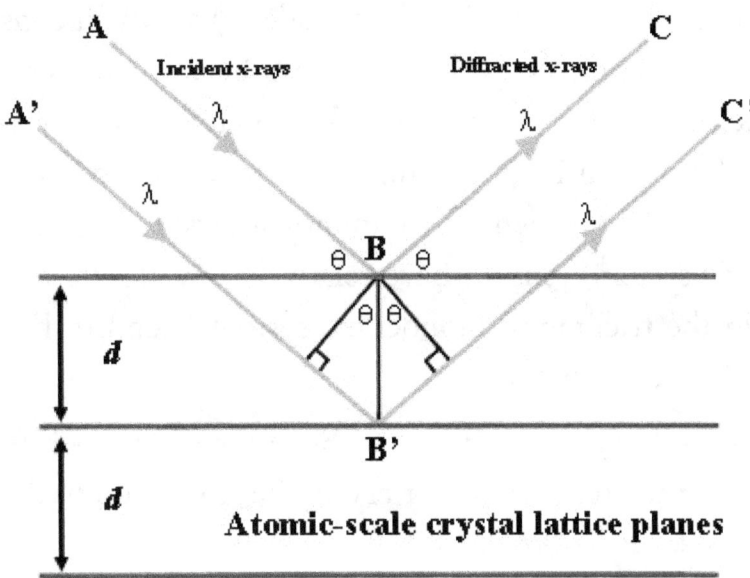

Figure: 3.1 Bragg's Law reflection. The diffracted X-rays exhibit constructive interference when the distance between paths ABC and A'B'C' differs by an integer number of wavelengths (λ).

When a crystal is bombarded with X-rays of a fixed wavelength (similar to spacing of the atomic-scale crystal lattice planes) and at certain incident angles, intense reflected X-rays are produced when the wavelengths of the scattered X-rays interfere constructively. In order for the waves to interfere constructively, the differences in the travel path must be equal to integer multiples of the wavelength. When this constructive interference occurs, a diffracted beam of X-rays will leave the crystal at an angle equal to that of the incident beam. To illustrate this feature, let us consider a crystal with crystal lattice planar distances d. Where the travel path length difference between the ray paths ABC and A'B'C' is an integer multiple of the wavelength, constructive interference will occur for a combination of that specific wavelength, crystal lattice planar spacing and angle of incidence (θ). Each rational plane of atoms in a crystal will undergo refraction at a single, unique angle (for X-rays of a fixed wavelength).

The general relationship between the wavelength of the incident X-rays, angle of incidence and spacing between the crystal lattice planes of atoms is known as Bragg's law, it is expressed as:

$$n \lambda = 2d \sin\theta \qquad\qquad \rightarrow 3.1$$

Where, the integer 'n' is the order of reflection, λ is the wavelength of the incident X-rays, d is the interplanar spacing of the crystal and 'θ' is the angle of incidence.

In X-ray diffraction (XRD) the interplanar spacing (d-spacing) of a crystal is used for identification and characterization purposes. In this case, the wavelength of the incident X-ray is known and measurement is made of the incident angle (θ) at which constructive interference occurs. Solving Bragg's Equation gives the d-spacing between the crystal lattice planes of atoms that produce the constructive interference. A given unknown crystal is expected to have many rational planes of atoms in its structure; therefore, the collection of "reflections" of all the planes can be used to uniquely identify an unknown crystal. In general, crystals with high symmetry (e.g. isometric system) tend to have relatively few atomic planes, whereas crystals with low symmetry (in the triclinic or monoclinic systems) tend to have a large number of possible atomic planes in their structures.

In X-ray diffraction and crystallography the Scherrer equation [200] is a formula that relates the size of sub-micrometer particles or crystallites in a solid to the peak broadening in a diffraction pattern.

The equation is given by

$$d = 2R = \frac{K\lambda}{\beta \cos\theta} \qquad\qquad \rightarrow 3.2$$

Where,

- d is the mean size of the ordered (crystalline) domains, which may be smaller or equal to the grain size;
- K is a dimensionless shape factor, with a value close to unity. K has a typical value of 0.89, but varies with the actual shape of the crystallite;
- ë is the wavelength of X–ray used;
- β is the line broadening at half the maximum intensity, i.e. full width half maximum (FWHM);
- θ is the Bragg angle.

The Scherrer equation is applied to determine size under the limitations to nano-scale particles only. It cannot be applied to grains larger than about 0.1 to 0.2 μm. It is to be noted that the Scherrer formula provides a lower bound on the particle size. The reason for this is that a variety of factors can contribute to the width of a diffraction peak besides instrumental effects and crystallite size; the most important of these are usually inhomogeneous strain and crystal lattice imperfections.

If all other contributions to the peak width were zero, then the peak width would be determined solely by the crystallite size and the Scherrer formula would apply. If the other contributions to the width are non-zero, then the crystallite size can be larger than that predicted by the Scherrer formula, with the "extra" peak width coming from the other factors.

The as-synthesized powder samples thus obtained were characterized by powder x-ray diffraction (XRD) (Philips 1730, Cu-$K\alpha$ radiation, l =1.54 Å). The XRD experiments were carried out by utilizing the facility available at the department of physics, Tezpur University.

3.2 UV-Visible spectroscopy

For the last few years and over this period the ultraviolet and visible spectrometers have been in general use and nowadays it is the most important analytical instrument in the modern day laboratory. In many applications, UV-Visible spectrometry has been widely used for its simplicity, versatility, speed, accuracy and cost-effectiveness.

Many molecules absorb ultraviolet or visible light. The absorbance of a solution increases as attenuation of the beam increases. Absorbance (**A**) is directly proportional to the path length, **b**, and the concentration, **c**, of the absorbing species. *Beers Law* [201] states that,

$$A = ebc,$$

Where **e** is a constant of proportionality, called the absorptivity. The figure: 3.2 shows a beam of monochromatic radiation of radiant power P_o, directed at a sample solution (S). Absorption takes place and the beam of radiation leaving the sample has radiant power P.

The amount of radiation absorbed may be measured as:

Transmittance, T =P/p_o

%Transmittance, %T=100T

$$\text{Absorbance, A} = \log_{10}(P/p_o) = 2 - \log_{10}(\%T) \qquad\qquad \rightarrow 3.3$$

Equation: 3.3 allows one to calculate absorbance from the transmittance data.

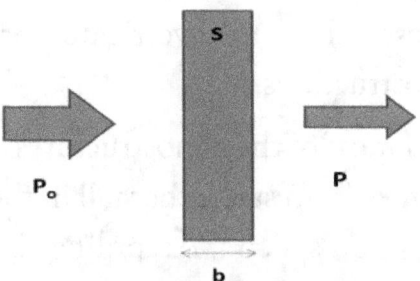

Figure 3.2: Change of radiant power due to absorption by sample solution.

Different molecules absorb radiation of different wavelengths. An absorption spectrum will show a number of absorption bands corresponding to structural groups within the molecule.

When white light falls upon a sample, the light may be totally reflected, in which case the substance appears white or the light may be totally absorbed, in which case the substance will appear black. If, however, only a portion of the light is absorbed and the balance is reflected, the colour of the sample is determined by the reflected light. Thus, if violet is absorbed, the sample appears yellow-green and if yellow is absorbed, the sample appears blue. The colours are described as complementary. However, many substances which appear colourless do have absorption spectra. In this instance, the absorption will take place in the infra-red or ultraviolet and not in the visible region. Table: 2 in the appendix illustrates the relationship between light absorption and colour.

A close relationship exists between the colour of a substance and its electronic structure. A molecule or ion will exhibit absorption in the visible or ultraviolet region when radiation causes an electronic transition within its structure. Thus, the absorption of light by a sample in the ultraviolet or visible region is accompanied by a change in the electronic state of the molecules in the sample. The energy supplied by the light will promote electrons from their ground state orbitals to higher energy, excited state orbitals or antibonding orbitals. In semiconductor material this corresponds to a transition from the valance band to conduction band edge. This characteristic absorption peak is used to study semiconductor materials in the bulk form or in the nano resigm.

In semiconductor nanostructures, the band gap enhancement is observed with decreasing cluster size due to quantum confinement effect. For this effect, a blue shift in the absorption edge is observed with decreasing crystalline/cluster size. The band gap energy of nanostructures can be estimated from UV-Visible study by using Bruss equation [202]

The energy of the nanocrystallite (E_{gn}) can be calculated from the UV-Vis absorption spectra using the relation hc/l, where l represent the corresponding absorption edge.

3.3 Photoluminescence spectroscopy

Beside UV-Visible spectroscopy, photoluminescence (PL) experiments are most widely used spectroscopic tool among the researchers to investigate optical transition in nanostructures especially in semiconductor nanostructures.

The emission characteristics of most of the nanostructures consist of a single-broad emission band, which is symmetric and comes from states that fall in the nanostructure's band gap. These states are not detectable in absorption spectra. According to L. Brus [202], the luminescence characteristics depend upon the nature of the semiconductors, the physical dimension as well as the chemical environment and the luminescence property can be manipulated in useful ways. Luminescence is the general term used to describe the emission of radiation from a solid when it is excited with some form of energy. When excitation arises from the absorption of photons, the phenomenon is known as photoluminescence. Whatever be the form of energy input, the final stage in the process is an electronic transition between two energy states E_1 and E_2 ($E_2 > E_1$), with the emission of radiation of wavelength λ where,

$$hc / \lambda = E_2 - E_1 \qquad \rightarrow 3.4$$

h and c being the Planck's constant and velocity of light respectively.

According to Stoke's law, the fundamental law of luminescence, the wavelength of emitted light is generally equal to or longer than that of the exciting light (i.e., of equal or less energy). This difference in wavelength is caused by a transformation of the exciting light, to a greater or lesser extent, to non-radiating vibration energy of atoms or ions. In rare instances e.g. when intense irradiation of laser beam is used or when sufficient thermal energy contributes to the electron excitation process—the emitted light can be of shorter wavelength than the exciting light (anti-Stokes radiation).

Usually, luminescence can be classified in two types: flouoroscence and phosphorescence. One can distinguish them depending on the duration of the emission, The fluorescence is an instantaneous process whereas in phosphorescence, the presence of vacant lattice sites or other impurities, lattice defects, and/or irregularities in the host lattice, provide unoccupied states (traps) and delay the luminescence by detaining (trapping) the charge carriers (electrons/holes) before their radiative recombination with the luminescent centres.

Depending on the nature of the ground and the excited states, the photoluminescence can be divided into two types. Singlet and triplet are the two different kinds of excited states. In a singlet state, the electron in the higher-energy orbital has the opposite spin orientation as the second electron in the lower orbital. These two electrons are said to be paired. On the other hand, in a triplet state these electrons are unpaired, i.e. their spins have the same orientation. An excited singlet state does not require an electron to change its spin orientation during the return to ground state. Contrary to this, a change of spin orientation is required for a triplet state to return to the ground state.

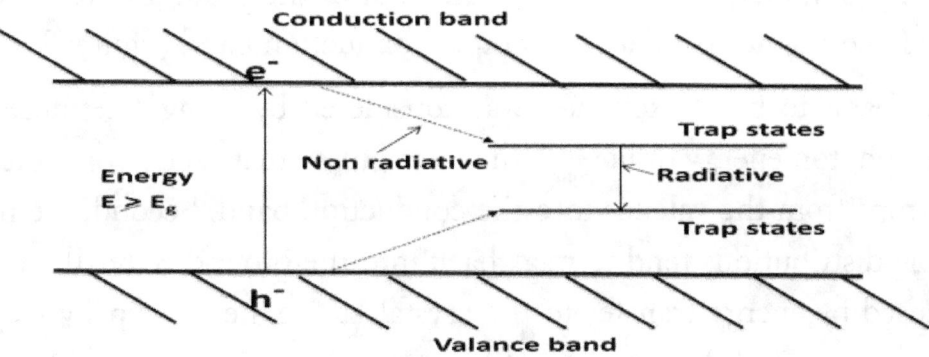

Figure 3.3: Radiative and non radiative emission and trap states.

In photoluminescence spectroscopy, photons with energy greater than the band gap of the semiconductor material studied are directed onto the surface of the material. The incident monochromatic photon beam is partially reflected, absorbed, and transmitted by the material being probed. The absorbed photons create electron-hole pairs in the semiconductor. The electrons are excited to the conduction band, or to the energy states within the gap. Moreover, electrons can lose part of their energy and transfer from the conduction band to energy levels within the gap. Photons produced as a result of the various recombinations of electrons and holes are emitted from the sample surface and it is the resulting photon emission spectrum (PL spectrum). The photon energies reflect the variety of energy states that are present in the semiconductor. In PL spectra, a direct conduction band-to-valence band recombination is rarely observed. Even if direct band-to-band recombinations occur, the crystal will strongly reabsorb the photons emitted. Therefore, in PL spectra, recombination processes are observed with emission energies less than E_g. These processes include excitonic recombinations and indirect transitions, which involve the trapping of electrons (or holes) by impurities. The nature of the quantum dot surface is critical for photoluminescence experiments. The influence of the surface on photoluminescence can be understood in terms of the trap states described in figure: 3.3[203]. The created electron–hole pair may recombine immediately to produce light (radiative recombination).

These trap states are created due to defects, such as vacancies, local lattice mismatches, dangling bonds, or adsorbents at the surface. The excited electron or hole can be trapped by these local energy minima states and become less available for the radiative recombination of luminescence. Radiative recombination of the trapped charge carriers then produces luminescence that is substantially redshifted from the absorbed light. Surface passivation is a well-known phenomenon that decreases the possibility of charge carriers residing in traps.

To have a better understanding regarding the PL experiments with semiconductors we can have a look through the simple description presented by Hanneewald, K., et al. [204].

A typical luminescence experiment in semiconductors can be realized in three stages, as presented in figure: 3.4; first, the sample is excited out of the ground state which is described by a completely filled valence band and an empty conduction band (figure: 3.4a).

Here, optical band-to-band excitation was considered by using a femtosecond (fs) laser pulse with mean photon energy of $\hbar\omega_{pump}$. The laser pulse creates electron-hole pairs due to a transfer of electrons from the valence into the conduction band. Second, the nonequilibrium electron and hole distributions tend to relax back into the ground state. The initial intraband relaxation is caused by energy transfer to the crystal lattice, i.e., a step-by-step excitation of lattice vibrations (figure: 3.4b), which are at low temperatures primarily longitudinal optical (LO) phonons in polar semiconductors such as Gallium Arsenide (GaAs). Finally, the electron-hole pairs recombine under emission of light which is the photoluminescence process (figure: 3.4c). Due to the attractive Coulomb interaction between the charge carriers, the emission spectrum does not only contain contributions from states at or above the fundamental energy gap E_{gap} but also sharp discrete lines just below E_{gap} which originate from bound excitonic states.

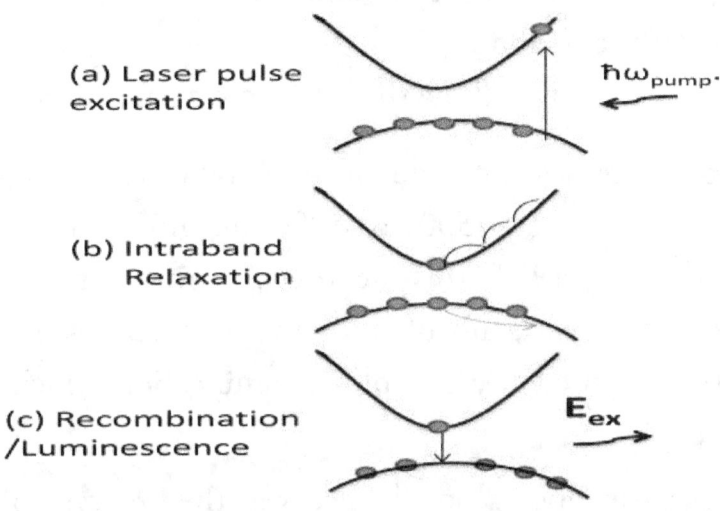

Figure 3.4: Three stages of a typical Luminescence experiment

3.4 Transmission Electron Microscope (TEM)

Nowadays, Transmission electron microscopy (TEM) is the most important tool for the researchers, particularly in the field of nanotechnology, since, by utilizing high resolution TEM (HRTEM) it is possible to detect the positions of atoms within a material.

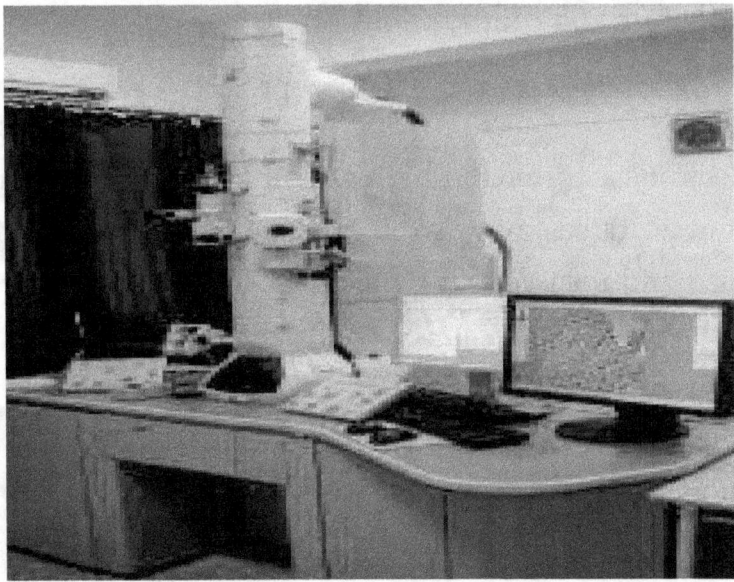

Figure 3.5: Transmission Electron microscopy at SAIF, Nehu, Shillong.

Electron Microscopes are scientific instruments that use a beam of highly energetic electrons to examine objects on a very fine scale. This examination can yield information regarding:

- The topography (surface features of an object),
- Morphology (shape and size of the particles making up the object),
- Composition (the elements and compounds that the object is composed of and the relative amounts of them) and
- Crystallographic information (how the atoms are arranged in the object).

Electron Microscopes were developed due to the limitations of Light Microscopes which are limited by the physics of light to 500x or 1000x magnification and a resolution of 0.2 micrometers [205]. In the early 1930's this theoretical limit had been reached and there was a scientific desire to see the fine details of the interior structures of organic cells (nucleus, mitochondria...etc.). This required 10,000x plus magnification which was just not possible using Light Microscopes.

The Transmission Electron Microscope (TEM) was the first type of Electron Microscope to be developed and is patterned exactly on the Light Transmission Microscope except that a focused beam of electrons is used instead of light to "see through" the specimen. It was developed by Max Knoll and Ernst Ruska in Germany in 1931. The first Scanning Electron Microscope (SEM) debuted in 1942 with the first commercial instruments around 1965. Its late development was due to the electronics involved in "scanning" the beam of electrons across the sample. Electron Microscopes (EMs) function exactly as their optical counterparts except that they use a focused beam of electrons instead of light to "image" the specimen and gain information as to its structure and composition.

The basic steps involved in all Electron Microscopes are:

- A stream of electrons is formed in high vacuum (by electron guns).
- This stream is accelerated towards the specimen (with a positive electrical potential) while is confined and focused using metal apertures and magnetic lenses into a thin, focused, monochromatic beam.
- The sample is irradiated by the beam and interactions occur inside the irradiated sample, affecting the electron beam.
- These interactions and effects are detected and transformed into an image.

The above steps are carried out in all EMs regardless of type.

Transimission Electron Microscopy (TEM) is a technique where an electron beam interacts and passes through a specimen. The electrons are emitted by a source and are focused and magnified by a system of magnetic lenses. The electron beam is confined by the two condenser lenses which also control the brightness of the beam, passes the condenser aperture and "hits" the sample surface. The electrons that are elastically scattered consist the transmitted beams, which pass through the objective lens. The objective lens forms the image display and the following apertures, the objective and selected area aperture are used to choose of the elastically scattered electrons that will form the image of the microscope. Finally, the beam goes to the magnifying system that is consisted of three lenses, the first and second intermediate lenses which control the magnification of the image and the projector lens. The formed image is obtained either on a fluorescent screen or in monitor or both and is printed on a photographic film. To detect the image, a high resolution phosphor may be coupled by means of a CCD (charge –couple device) camera. The image, thus detected by the CCD camera can be interfaced to a PC for data actuation and image display [206].

We have performed TEM investigations of my samples at SAIF, Nehu, Shillong by utilizing a **JEOL** model **1200 EX** transmission electron microscope (figure: 3.5). The detailed specification is given in the appendix in table: 5.

3.5 Scanning Probe Microscope (SPM)

The basic principle of this microscope [207] is to measure forces or to measure interactions between a sharp probing tip and sample surface led to the creation of a variety of other scanning probe microscopes (SPM), such as:

- The magnetic force microscope (MFM),
- The dipping force microscope (DFM),
- The friction force microscope (FFM), and
- The electrostatic force microscope (EFM).

By these new developments the field became further subdivided. Concurrently, there is also an unifying tendency to combine different methods such as STM/AFM, AFM/MFM, AFM/FFM. This provides the unique opportunity to characterize a single nm-sized spot by a combination of methods and therefore gain more information than by the separate application of a single method.

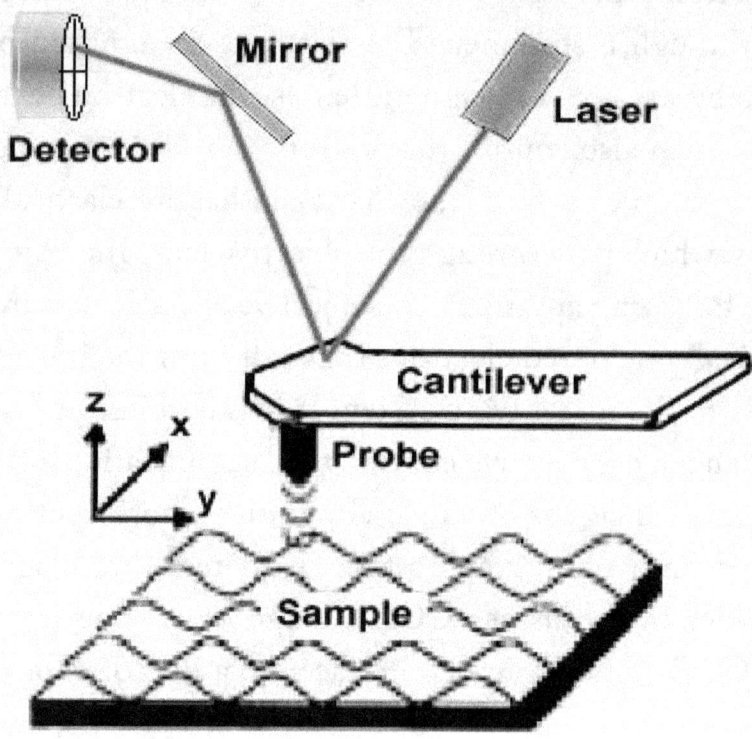

Figure 3.6: Basic principle of AFM/AMF, A sharp tip is mounted on a cantilever-type spring. The force between the tip and the sample causes cantilever deflections which are monitored by using a deflection sensor.

In probing force microscopy the probing tip is attached to a cantilever-type spring. In response to the force between tip and sample the cantilever is deflected. Images are obtained by scanning the sample relative to the probing tip and digitizing the deflection of the lever or the z-movement of the piezo as a function of the lateral position x, y. Typical spring constants are between 0.001 to 100 N/m and motions from microns to ~0.1 Å are measured by the deflection sensor (figure: 3.6). Typical forces between probing tip and sample range from 10^{-11} to 10^{-6} N. One can realize this, from the comparison of the interaction between two covalently bonded atoms, that is of the order of 10^{-9}N at separations of ~ Å. Thus, non-destructive imaging is possible with these small forces. Force regimes can be classified as: (a) Contact mode and (b) Non-contact mode. When the microscope is operated in non-contact mode at tip-sample separations of 10 to 100 nm, forces, such as Vander-Waals, electrostatic, magnetic or capillary forces, can be sensed and get information regarding surface topography, distributions of charges, magnetic domain wall structure or liquid film distribution. At smaller separations of the order of Å the probing tip is in contact with the sample. In this mode, ionic repulsion forces allow the surface topography to be traced with high resolution. Under best conditions atomic resolution is achieved. In addition, frictional forces and elastic or plastic deformations can be detected under appropriate conditions.

In my investigations we have utilized the unique combination probing force microscopy (AFM/MFM) to obtain surface morphology and magnetic behaviour of the Cr:ZnS and Mn:ZnS systems. For MFM measurements, the tip used in AFM measurements has been changed with a magnetic tip.

3.6 Electron Dispersive X-ray spectrometry (EDX)

An EDX instrument can be attached to Scanning electron microscope (SEM) to provide supplementary information. By utilizing EDX investigations, one can get information about composition of individual crystals or features. EDS makes use of the X-ray spectrum emitted by a solid sample bombarded with a focused beam of electrons to obtain a localized chemical analysis. **Qualitative analysis** involves the identification of the lines in the spectrum and is fairly straightforward owing to the simplicity of X-ray spectra. **Quantitative analysis** (determination of the concentrations of the elements present) entails measuring line intensities for each element in the sample and for the same elements in calibration Standards of known composition. As the SEM electron beam strikes the sample surface, X-rays are produced. An X-ray photon impinging on the surface of the EDX detector produces electron hole pairs which are detected as a single pulse by the liquid nitrogen cooled pre amplifier. The pulse energy is determined by the X-ray energy which in turn is determined by the element being determined. The EDX analyser produces a spectrum of the elements present in targeted areas of the samples allowing detectable elements to be quantified or mapped.

3.7 Fourier Transform infrared spectroscopy (FTIR)

FT-IR Spectroscopy is the abbreviation of "Fourier Transform Infrared Spectroscopy". This is the analytical technique developed in 1970s to qualify and quantify compounds utilizing infrared absorption of molecules.

In organic chemistry, Infrared spectroscopy is an important technique. It offers easy way to identify the presence of certain functional groups in a molecule. Moreover, one can use the unique collection of absorption bands to confirm the identity of a pure compound or to detect the presence of specific impurities. Infrared (IR) lights are electromagnetic radiation with their wavelength longer than those of visible lights, measuring from the nominal edge of visible red light at 0.74 micrometers, and extending conventionally to 300 micrometers. Absorption occurs when the energy of the beam of light (photons) are transferred to the molecule. The molecules get excited and move to a higher energy state. The energy transfer takes place in the form of electron ring shifts, molecular bond vibrations, rotations and translations. FTIR is mostly concerned with vibrations and stretching.

An FTIR (Fourier Transform Infra Red) is a method of obtaining infrared spectra by first collecting an interferogram of a sample signal using an interferometer, and then performing a Fourier Transform (FT) on the interferogram to obtain the spectrum. An FTIR Spectrometer collects and digitizes the interferogram, performs the FT function, and displays the spectrum.

An FT-IR is typically based on The Michelson Interferometer. The interferometer consists of a Beam-splitter, a fixed mirror, and a mirror that translates back and forth, very precisely. The beam splitter is made of a special material that transmits half of the radiation striking it and reflects the other half. Radiation from the source strikes the beam splitter and separates into two beams. One beam is transmitted through the beam splitter to the fixed mirror and the second is reflected off the beam splitter to the moving mirror. The fixed and moving mirrors reflect the radiation back to the beam-splitter. Again, half of this reflected radiation is transmitted and half is reflected at the beam splitter, resulting in one beam passing to the detector and the second back to the source.

3.8 Superconducting quantum interference device (SQUID)

Figure 3.7: SQUID facility at DAE consortium, Indore.

Superconducting interference device (SQUID) is a very useful tool that can measure extremely low magnetic field. It is potentially more sensitive than a normal vibration magnetometer. It is sensitive enough to measure magnetic fields as low as 5×10^{-18} T [208], for which researchers usually prefer SQUID measurements for characterization of magnetic samples. SQUID offers many advantages like high sensitivity, low noise, less error etc., for which it plays a vital role among the research groups particularly in the field of magnetic semiconductor nanostrures.

SQUID is basically based on the superconducting loops containing Josephson junctions. The phenomenon of supercurrent (current flows indefinitely long without any voltage applied across a device is known as a Josephson junction (JJ), that consists of two superconductors coupled by a weak link. It is named after the British physicist Brian David Josephson, who predicted in 1962 the mathematical relationships for the current and voltage across the weak link [209,210].

A superconducting quantum interference device (SQUID) is a mechanism used to measure extremely weak signals, such as subtle changes in the human body's electromagnetic energy field. Using Josephson junction, a SQUID can detect a change of energy as much as 100 billion times weaker than the electromagnetic energy that moves a compass needle. A Josephson junction is made up of two superconductors, separated by an insulating layer so thin that electrons can pass through. A SQUID consists of tiny loops of superconductors employing Josephson junctions to achieve superposition: each electron moves simultaneously in both directions. Because the current is moving in two opposite directions, the electrons have the ability to perform as qubits (that theoretically could be used to enable quantum computing). SQUIDs have been used for a variety of testing purposes that demand extreme sensitivity, including engineering, medical, and geological equipment. Because they measure changes in a magnetic field with such sensitivity, they do not have to come in contact with a system that they are testing. SQUIDs are usually made of either a lead alloy (with 10% gold or indium) and/or niobium, often consisting of the tunnel barrier sandwiched between a base electrode of niobium and the top electrode of lead alloy. A radio frequency (RF) SQUID is made up of one Josephson junction, which is mounted on a superconducting ring. An oscillating current is applied to an external circuit, whose voltage changes as an effect of the interaction between it and the ring. The magnetic flux is then measured. A direct current (DC) SQUID, which is much more sensitive, consists of two Josephson junctions employed in parallel so that electrons tunnelling through the junctions demonstrate quantum interference, dependent upon the strength of the magnetic field within a loop. DC SQUIDs demonstrate resistance in response to even tiny variations in a magnetic field, which is the capacity that enables detection of such minute changes.

We have utilized SQUID facility at UGC DAE Consortium, Indore to investigate magnetic properties of my samples within temperature range 10K to 300K. The magnetometer is shown in figure: 3.7.

3.9 Significant observations

By utilizing different characterization techniques the structural, optical and magnetic properties of TM doped ZnO (TM=Mn, Co and Ni) and Mn, Cr doped ZnS was investigated systematically.

- **Structural** study of as-synthesised TM doped ZnO, ZnS samples was carried out by utilizing XRD, FTIR, HRTEM, and EDX. The HRTEM study confirmed the formation of elongated nanostructures for bare and TM doped ZnO, while most of the nanostructures were found to be spherical in nature for TM doped ZnS samples. In case of Cr:ZnS development of fractal pattern were detected at low resolution TEM study. XRD patterns exhibited the wurtzite nature of all TM doped and bare ZnO samples, while cubic crystalline structure for TM:ZnS system. Surface morphology and existence of magnetic domains of TM:ZnS system were depicted from AFM and MFM experiments.

- **Optical** properties of the samples were explored from PL and UV-Visible study. PL intensities for doped and bare ZnO and ZnS samples were found to exhibit typical nature of DMS. Band gap study of the sample was studied through UV-Vis measurements.

- **Magnetic** properties of the samples were explored with SQUID within temperature range 4K to 300K. M~T response were studied for both ZFC and FC. M~H responses were measured at both low temperature (10K) as well as room temperature (300K). Existence of room temperature ferromagnetism was confirmed through SQUID measurements for Co, Mn and Ni doped ZnO samples.

ZnO Based DMS Nanostructures

The study of nanostructures with controlled morphology, shapes and size is essential for developing materials with novel properties and tailorable functions. Different types of semiconducting nanomaterials have attracted a large group of scientific community because of their exceptional properties, which are different from bulk materials [211, 212]. Zinc oxide (ZnO), one of the very important and versatile semiconductors with direct band gap of gap of ~3.37 eV and a large exciton binding energy of ~60 meV at room temperature (RT) is a promising candidate for functional components of devices. Selective doping with transition metal ions into ZnO lattice host is capable of tailoring its physical properties. In this chapter, the structural as well as optical properties of bare ZnO and transition metal (TM:Mn, Co, Ni, Cu) doped ZnO is highlighted. We have fabricated TM doped ZnO and bare ZnO by adopting solid state chemical reaction route which has been described in chapter-2. The concentration of TM were varied from 1% to 5%, accordingly they were indexed as TM-1, TM-3 and TM-5 (TM: Mn, Co, Ni, Cu). At the end of synthesis, all samples were obtained in the form of powder. Structural and optical investigations of the as fabricated samples were carried out by utilizing XRD, EDS, HRTEM, FTIR, PL and UV-Vis spectroscopy. The change of structural and optical properties due to different TM-doping is one of the major directions of my investigation

4.1 Mn doped ZnO

We have fabricated undoped ZnO and Mn:ZnO nanostructures for 1-3% Mn-doping cases, accordingly the samples were indexed as ZnO, Mn-1, Mn-3 and Mn-5. TEM analysis confirmed the formation of bare ZnO and Mn:ZnO nanorods having average aspect ratio ~3.7 and ~3.3 respectively.

(i) X-ray diffraction study

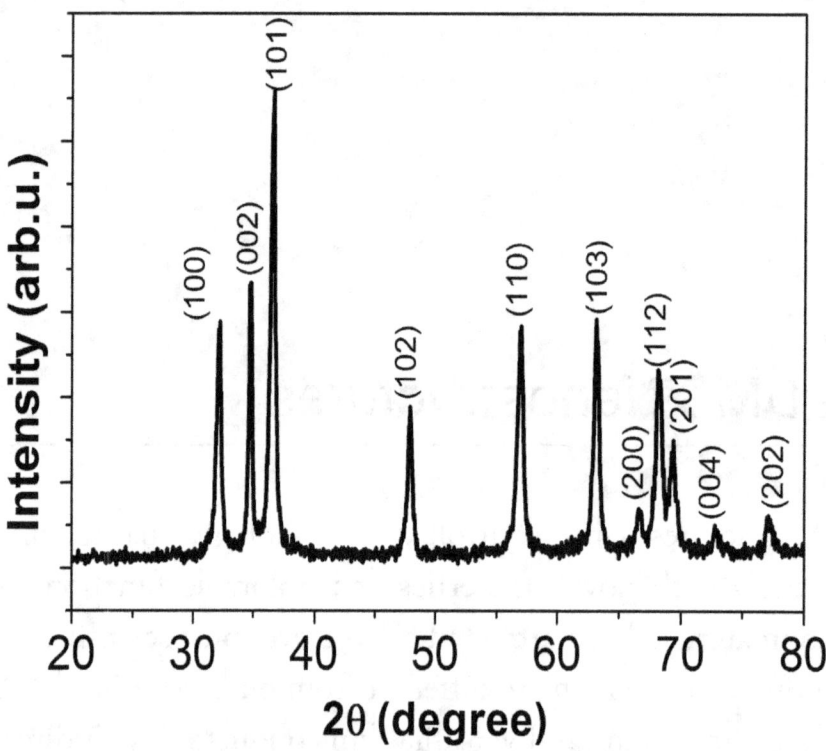

Figure 4.1: XRD pattern of undoped ZnO nanorods showing prominent diffraction peaks which can be perfectly indexed to the hexagonal wurtzite structure of ZnO (JCPDS card no. 36-145).

The phase and purity of the as prepared samples were determined by X-ray powder diffraction pattern. For identification of structural phase and crystallographic orientation we have utilized Rigaku Miniflex diffractometer employing Cu-K$_\alpha$ radiation at 1.54 Å with a scanning rate of 0.02 deg/s. XRD measurements have showed that the undoped and Mn doped ZnO crystallize into hexagonal wurtzite structures. The XRD pattern for undoped ZnO as presented in figure:4.1. shows prominent diffraction peaks which can be perfectly indexed to the hexagonal wurtzite structure of ZnO according to JCPDS data card no.36-145. The XRD data of the Mn:ZnO samples have ensured wurtzite structure of ZnO with a high degree of crystallinity. The comparative X-ray diffraction pattern for bare ZnO and Mn:ZnO system is shown in figure: 4.2. No extra phases e.g. MnO_2 etc. were observed for Mn-doped ZnO systems. Thus Mn is expected to occupy Zn-sites of the host ZnO lattice. Owing to lattice mismatch introduced by Mn-doping, there is a slight shifting of XRD peaks [figure: 4.3] for different Mn concentrations. The most intense (101) peak for bare ZnO sample was observed at a Bragg angle (2θ) of ~36.6⁰.

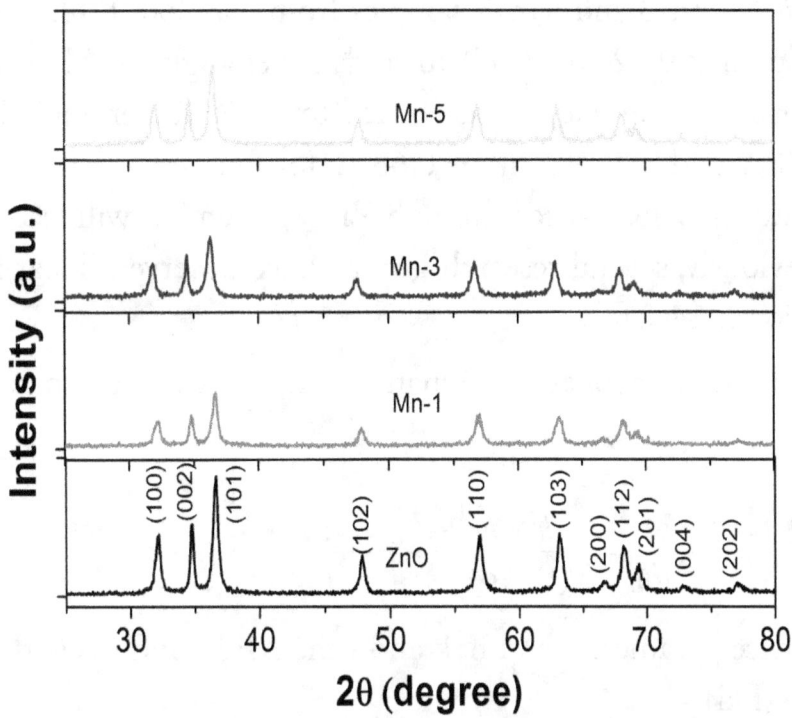

Figure 4.2: XRD patterns of Bare ZnO and Mn doped nanorods: Mn-1, Mn-2 and Mn-3.

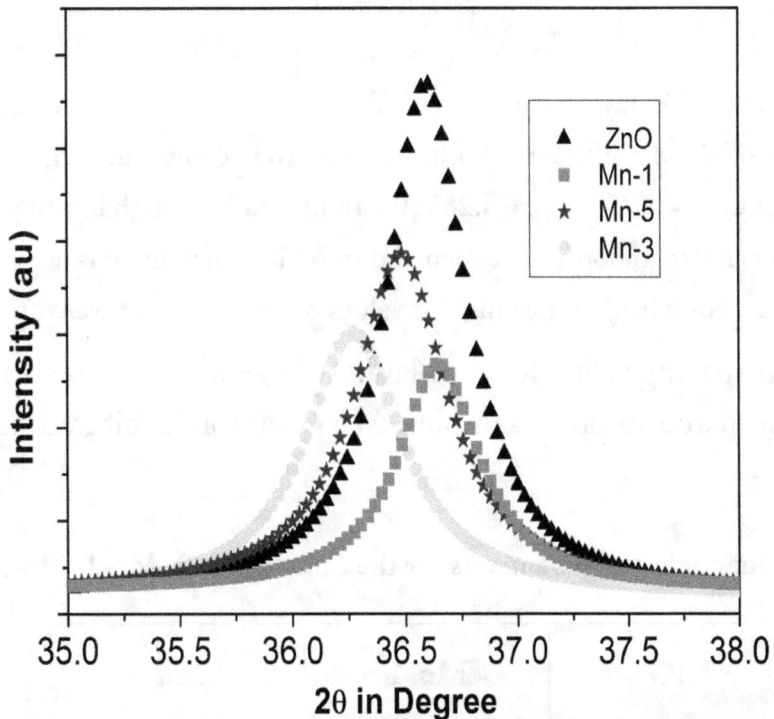

Figure 4.3: Lorentzian fit showing Shifting of XRD peaks for (101) plane of bare ZnO, Mn-1, Mn-2 and Mn-5 samples.

A close observation to detect the shifting of XRD peaks for the plane (101) was recorded from Lorentzian fit of the curves in between 35^0 and 38^0 is presented in figure: 4.3, which indicates clear shifting for Mn-1, Mn-3 and Mn-5 samples from bare ZnO. The shifting of peaks for (101) plane are measured as 0.32^0 and 0.11^0 towards lower angle for Mn-3 and Mn-5 samples respectively, while for Mn-1 sample, it was shifted towards higher angle by 0.05^0. The ionic radius of Zn^{2+} is 0.60 Å, and that of Mn^{2+} is 0.66 Å, for four-fold coordination [213]. Hence, Mn incorporation into the ZnO lattice through Zn replacement will lead to an expansion of the ZnO lattice. Previously, several research groups have observed linear increase of *a* and *c* with Mn concentration [214-216].

Lattice parameter for hexagonal ZnO nanostructures were estimated by utilizing the equation:

$$\frac{1}{d_{hkl}^2} = \frac{4}{3}\left(\frac{h^2 + hk + k^2}{a^2}\right) + \frac{l^2}{c^2} \qquad \rightarrow 4.1$$

Being a and c the lattice parameters and h, k and l the Millar indices and d_{hkl} the interplanar spacing for the plane (hkl).

This interplanar spacing can be calculated from the equation (3.1)

$$2d \sin\theta = n\lambda$$

Where ë is the wavelength of X-ray used, θ is the diffraction angle and n is the order of diffraction (n=1). The experimental results of 'a' and 'c' estimated for my samples were compared with their theoretical values for wurtzite phase of bulk ZnO (a = 3.250 A^0, c = 5.265 A^0). In my study, though lattice parameters get enhanced in Mn-doping cases compared to undoped specimen, the amount of increment is not substantial between Mn-doped samples (table: 4.1). For Mn-3 sample highest values of enhancement were observed.

Table: 4.2 shows 'd' spacing values for the planes (001) and (002) as estimated for all samples. We observed that compared to other samples, Mn-3 sample exhibited maximum value for 'd' spacing.

Table 4.1: Values of lattice parameters for the samples ZnO, Mn-1, Mn-3 and Mn-5

Lattice parameter	ZnO	Mn-1	Mn-3	Mn-5
a (A^0)	3.2173	3.2232	3.2474	3.2265
c(A^0)	5.152	5.1747	5.1995	5.1701

Table 4.2: Values of lattice lattice spacing 'd' for the planes (100) and (002) for the samples ZnO, Mn-1, Mn-3 and Mn-5

(h k l)	ZnO (A⁰)	Mn-1 (A⁰)	Mn-3 (A⁰)	Mn-5 (A⁰)
(1 0 0)	2.786	2.791	2.812	2.794
(0 0 2)	2.576	2.587	2.599	2.585

(a) Size determination

Size determination of the nanostructures so formed was performed by utilizing three methods (i) Scherrer formula, (ii) Williamson-Hall Plot and (iii) TEM analysis. The Scherrer formula is described in chapter -3. It is the simplest approach to calculate size from XRD line broadening. The Scherrer equation (3.2) is given by:

$$d = 2R = \frac{0.98\,\lambda}{\beta\,\cos\theta}$$

Where d is the crystallite size, ë the wavelength of X-ray diffraction, β the full width at half maximum (FWHM) of the diffraction peak (in radians), θ is the maximum scatter angle.

For size determination of the nanocrystals, usually, the most intense peak has been selected from the XRD spectra. In my case the most intense peak is obtained from the reflection of (101) plane for all the samples.

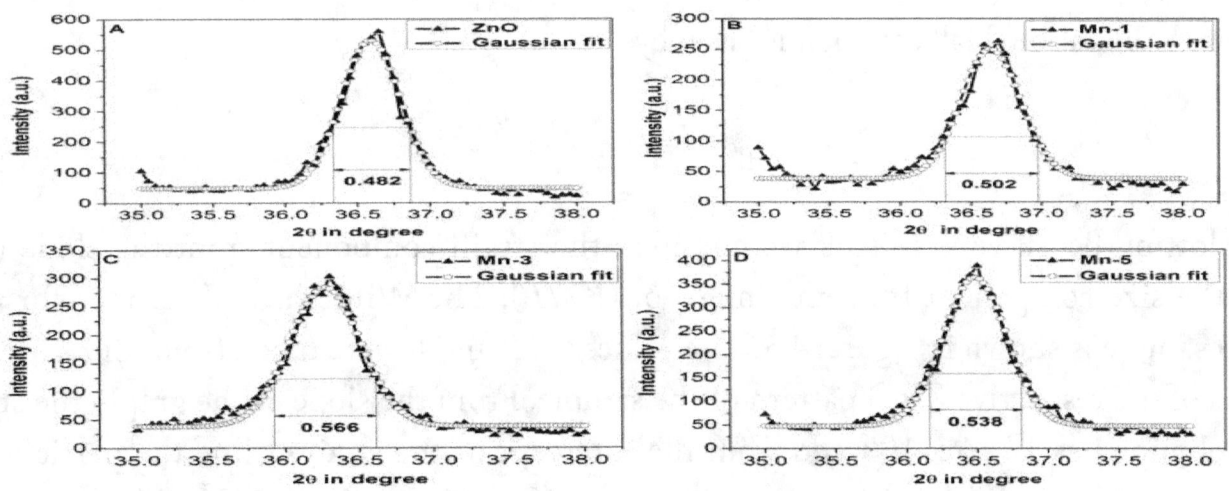

Figure 4.4: Individual Gaussian fit for the peak broadening reflected by the plane (101) for the samples (A) ZnO, (B) Mn-1, (C) Mn-3 and (D) Mn-5

The individual Gaussian fit for the peak broadening reflected by the plane (101) for the samples (A) ZnO, (B) Mn-1, (C) Mn-3 and (D) Mn-5 is presented in figure:4.4.

The sample Mn-3 showed maximum broadening of ~0.57⁰ (figure: 4.4C). The lowest peak broadening was obtained for ZnO sample (figure: 4.4A). The variation of Full width half maxima (FWHM) corresponding to different peak positions with crystallite sizes is given in the table: 4.3. The average crystallite size as obtained from individual XRD line broadening and using Scherrer formula [217] for the ZnO, Mn-1, Mn-3 and Mn-5 were 17.15, 16.47, 14.61 and 15.37 nm; respectively. A decrease in the average particle size with increasing Mn content has been reported earlier [218, 219]. Similar result was observed in my analysis.

Table 4.3: Varaiation of FWHM, peak position of (101) plane and crystallite size for all samples.

	FWHM (In degree)	2θ value (In degree)	Crystallite size (nm)
ZnO	0.482	36.59	17.15
Mn-1	0.502	36.64	16.47
Mn-3	0.566	36.27	14.61
Mn-5	0.538	36.48	15.37

(b) Williamson Hall plot

Calculation of size and strain were also performed by plotting Williamson-Hall plots for the samples. The Williamson-Hall plot gives information regarding the size and strain of the nanostructures.

The Williamson-Hall equation is given by

$$\beta\cos\theta = C_\varepsilon\, 4\sin\theta + \frac{K\lambda}{L} \qquad\qquad \rightarrow 4.1$$

By plotting $\beta\cos\theta$ versus $4\sin\theta$ we obtained the strain component from the slope ($C\varepsilon$) and the size component from the intercept ($K\lambda/L$). The Williumson-Hall plot for pure ZnO sample is shown in figure: 4.5. The values for ß and θ were taken from four different peaks of the respective XRD pattern of the sample. From the slope of the graph, the strain was obtained as 14.46×10^{-2} and from the intercept of the straight line the particle size (L) was estimated as 12.82 nm by considering $K = 0.9$ and $\lambda = 1.54$ Å. The strain and particle size of other nanostructures so fabricated were calculated from the analysis of

Williamson-Hall plot and presented in the table. The respective values for particle strain and size are presented in the table 4.4. There is a linear increase in strain with increasing particle size. An anomaly regarding the particle size was observed in this case. This may be due to the fact that, Scherrer approach is based upon the assumption where strains and faulting are ignored for a small cubic crystal and the peak broadening is only due to the small size.

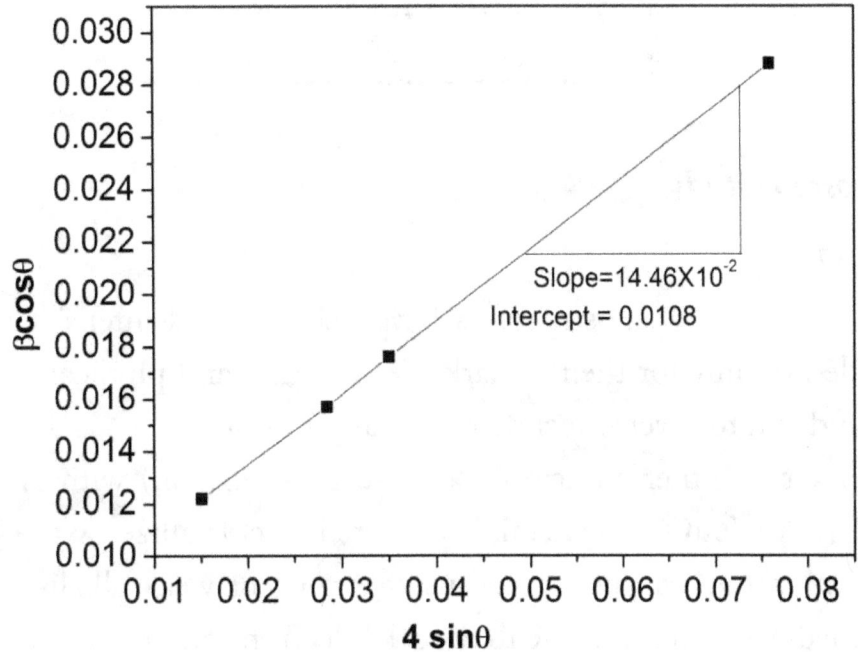

Figure 4.5: Williamson-Hall plot for bare ZnO sample.

However the Williamson-Hall method has many assumptions: its absolute values should not be taken too seriously but it can be a useful method if used in the relative sense; for example a study of many powder patterns of the same chemical compound, but synthesised under different conditions, might reveal trends in the crystallite size/strain which in turn can be related to the properties of the product.

Table 4.4: Strain and size of the nanostructures ZnO, Mn-1, Mn-3 and Mn-5 as obtained from their respective Williamson-Hall plots.

Sample	Strain	Size (nm)
ZnO	14.46×10^{-2}	12.82
Mn-1	36.96×10^{-2}	24.2
Mn-3	27.72×10^{-2}	18
Mn-5	8.69×10^{-2}	9.5

(ii) Electron microscopy study

(c) ZnO nanostructures

One-dimensional (1D) ZnO nanostructures have been studied intensively and extensively over the last decade not only for their remarkable chemical and physical properties, but also for their current and future diverse technological applications. ZnO is an amphoteric oxide (an oxide that can act as either an acid or as base in a reaction) with an isoelectric point value of about 9.5 [220]. ZnO, in general, is expected to crystallize by the hydrolysis of Zn salts in a basic solution that can be formed using strong or weak alkalis. Zn^{2+} is known to coordinate in tetrahedral complexes. Due to the $3d^{10}$ electron configuration, it is colourless and has zero crystal field stabilization energy. Depending on the given pH and temperature [221], Zn^{2+} is able to exist in a series of intermediates, and ZnO can be formed by the dehydration of these intermediates. Chemical reactions in aqueous systems are usually considered to be in a reversible equilibrium, and the driving force is the minimization of the free energy of the entire reaction system, which is the intrinsic nature of wet chemical methods [222]. Wurtzite structured ZnO grown along the c axis has high energy polar surfaces such as ± (0001) surfaces with alternating Zn^{2+}-terminated and O^{2-}-terminated surfaces [223]. So when a ZnO nucleus is newly formed, owing to the high energy of the polar surfaces, the incoming precursor molecules tend to favourably adsorb on the polar surfaces. However, after adsorption of one layer of precursor molecules, the polar surface transforms into another polar surface with inverted polarity. For instance, a Zn^{2+}-terminated surface changes into an O^{2-}-terminated surface, or vice versa. Such a process is repeated over time, leading to a fast growth along the ± [0001] directions, exposing the non-polar {1100} and {2110} surfaces to the solution. This is essentially how a 1D nanostructure is formed [224]. In my study, we have observed the formation of one dimensional nanostructures of as synthesised bare ZnO and transition metal

(Mn, Co, Cu) doped ZnO samples synthesised via solid state chemical reaction route. The elongated nanostructure was confirmed through transmission electron microscope (TEM) study.

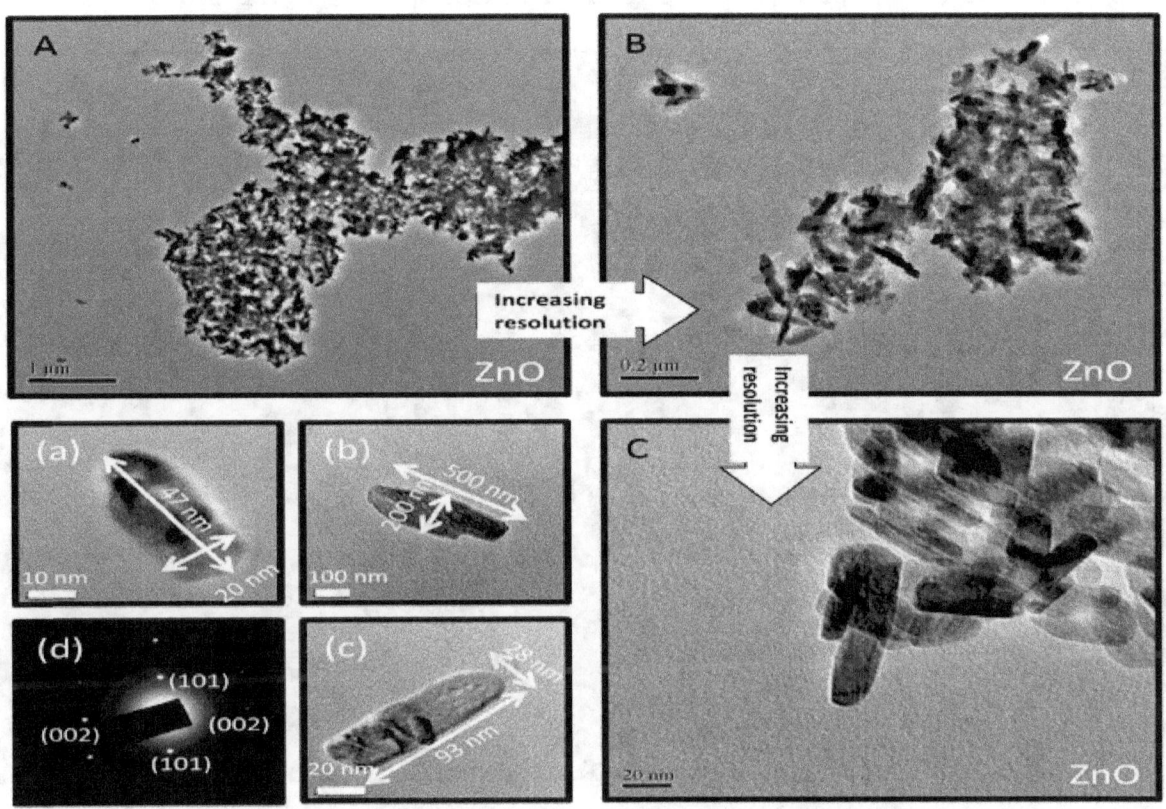

Figure 4.6: TEM image of the ZnO sample, with increasing resolution from figure: A to C. With high resolution TEM micrograph three isolated elongated nanostructures having different lengths and diameters, (a)47 nm, 20 nm; (b) 500 nm, 200 nm; and (c) 93 nm, 28 nm; respectively. (d) Selected area electron diffraction pattern (SAED) focussed on the isolated nanorod (c).

For TEM study, I have utilised JEM-2100 model with resolution 1.9Å to 1.4Å, accelerating voltage 60-200 KV in 50 steps having magnification 50 to 1500000 times.

During my TEM observations, focussing the electron beam on the ZnO cluster, the presence of one dimensional ZnO nanostructures was confirmed.

Various TEM images for the bare ZnO sample are presented in figure: 4.6. Changing the resolution from low to high with increasing magnification (from figure: 4.6A to 4.6C) we confirmed the formation of ZnO nanorods. During TEM investigation few ZnO nanorods were found to be isolated, three of them are presented in figure: 4.6(a) - 4.6(c). Interestingly, a knife like elongated structure with maximum length and diameter ~500 nm and ~200 nm respectively was observed in the TEM micrograph (figure: 4.6b). Estimated lengths and diameters of other two isolated rods as presented in figure: 4.6(a) and figure: 4.6(c) were 47 nm,

20 nm and 93 nm, 28 nm respectively. Uniformly distributed ZnO lattice points are observed in the selected area electron diffraction (SAED) pattern [figure: 4.6(d)]. All points correspond to the hexagonal wurtzite structure of ZnO. Four points within two inner rings are ascribed for the planes (101) and (002) corresponding to their lattice spacing 2.4Å and 2.6 Å.

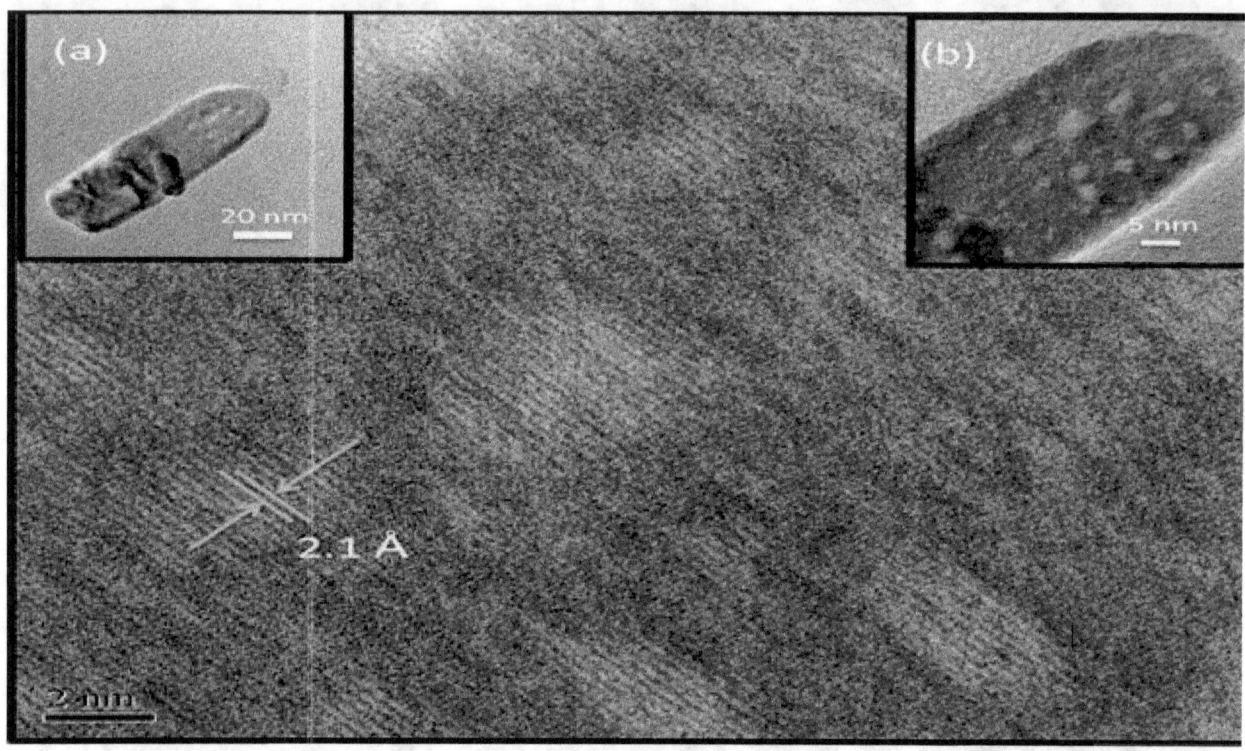

Figure 4.7: HRTEM image of ZnO nanorod, inset isolated nanorod on which the beam is focussed to get the HRTEM image (a) at low magnification with 20 nm scale, (b) at high magnification with 5 nm scale.

The HRTEM study was performed on the isolated nanorod as presented in figure: 4.7(a), the resolution was increased to get the pattern shown in the inset (b) of figure: 4.7. On further increasing the resolution, lattice planes with few defect states were observed. The lattice spacing was estimated to be ~2.1Å at a position where lattice fringes are clear enough, this lattice spacing 'd' corresponds to the plane (102) as calculated from the XRD pattern of the ZnO sample. Thus, the growth direction of small crystallite is identified to be along (102) plane. It has been noticed that the surface of the small crystallite exposed to the beam consists of several lattice distortion. This structural property may play important role to change the optical, magnetic and other properties of the sample.

(d) ZnO and Mn doped ZnO

The Ttransmission electron micrographs (TEM) of the ZnO, Mn-1, Mn-3 and Mn-5 samples give clear visual evidence on the formation of nanorods as presented in Figure:4.8.

Nanorods of different lengths and diameters with a random orientation can be seen from the micrographs. For systematic measurement of the lengths and diameters of the nanorods we have utilized Sun Microsystems, Inc. software "Java SE runtime environment (JRE) version 6" and "Java FX runtime verson 1". For ZnO sample we have selected 10 nanorods which are almost straight and isolated. The average length of the rods was found to be ~70 nm. The average diameter of the nanorods is measured as ~19 nm which is comparable with the average size predicted from XRD data. The average aspect ratio of the ZnO nanorods obtained as 3.7. We speculate that the most of the rods are composed of monocrystallites over a definite length. The average length, diameter, and aspect ratio for the samples as calculated from TEM image were shown in table: 4.5. Compared to bare ZnO sample a slight decrease in aspect ratio was observed for Mn doped samples but it was not substantial. Mn-3 sample yields nanorods with average length as 91 nm which is higher than any other sample.

Table 4.5: Lengths and diameters of ZnO nanorods (10 nos) as measured from TEM images (figure: 4.8A) by using standard software.

Sl. No.	1	2	3	4	5	6	7	8	9	10
Length (nm)	64	54	71	75	73	76	46	101	75	72
Diameter (nm)	11	15	18	25	20	18	24	21	15	24

Table 4.6: Average length, diameter, aspect ratio for the samples ZnO, Mn-1, Mn-3 and Mn-5 as obtained from TEM analysis.

Sample	Av. length L (nm)	Av. Diameter D (nm)	Aspect ratio (L/D)
ZnO	70.6	19	3.7
Mn-1	63.28	18.38	3.4
Mn-3	91	28	3.2
Mn-5	62	17.5	3.5

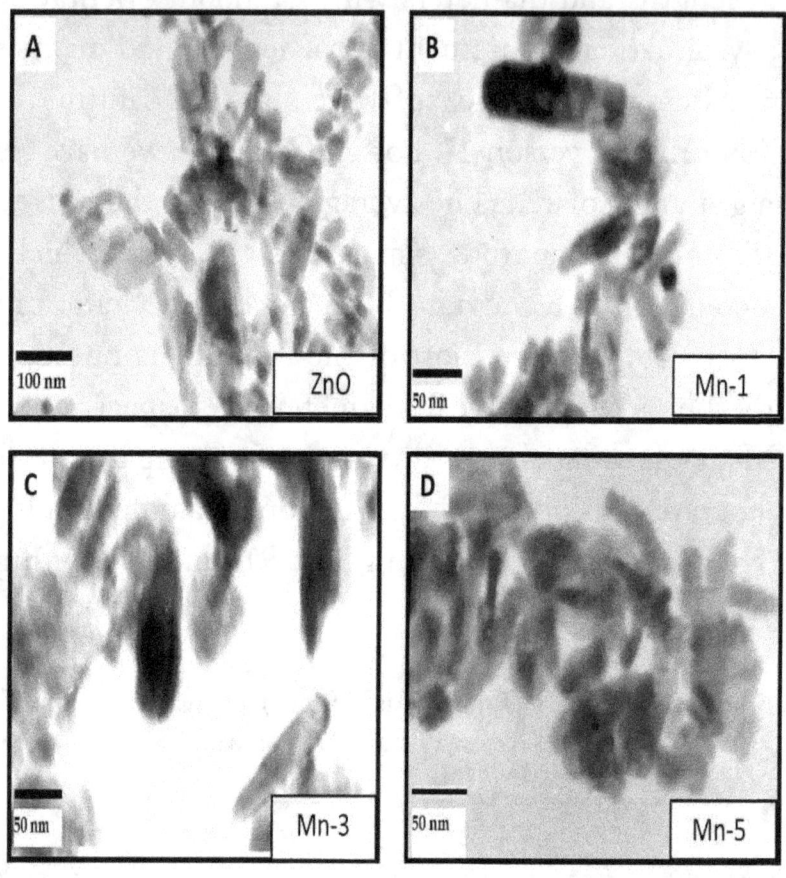

Figure 4.8: TEM image of the samples (A) ZnO, (B) Mn-1, (C) Mn-3, (D) Mn-5

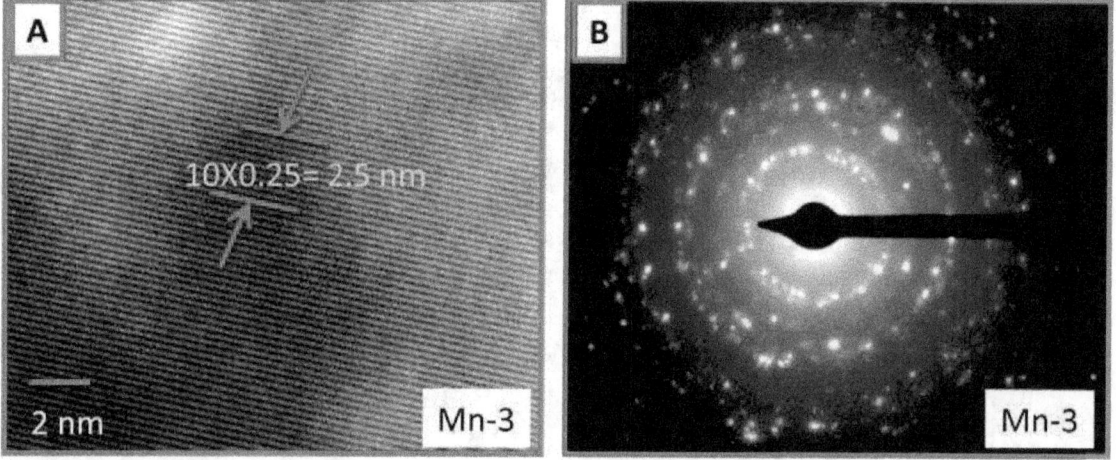

Figure 4.9: (A) HRTEM image collected for the sample Mn-3, (B) SAED pattern for the sample Mn-3.

High resolution transmission electron microscopy (HRTEM) image for the Mn-3 sample as presented in figure: 4.9(A) yields formation of high crystalline nature of the nanocrystals as prepared. The lattice spacing as measured from the image is 2.5 Å corresponding to the plane (101), it indicates the growth direction of the nanocrystal is along (101). Clear fringe pattern also reveals nonexistence of lattice distortion of the as fabricated Mn-3 sample. The selective area electron diffraction (SAED) pattern for the sample Mn-3 is presented in Figure: 4.9(B). It reveals diffused rings which correspond to the hexagonal wartzite structure of ZnO without any impurity phase. As observed from the pattern, we ascribe two inner rings as (100) and (002) planes for which the *d-* spacing is calculated to be 2.86 nm and 2.47 nm. The values correspond to the *d-* spacing parameters as calculated from XRD data (Table 4.2).

(iii) Fourier Transform Infrared spectroscopy study

An infrared spectrum is considered to be the fingerprint of a given specimen with absorption peaks corresponding to the frequencies of inter-atomic vibrations in the molecular system. Infrared spectrum can be considered as an important asset in the qualitative analysis of a given system, because no two different compounds can ever exhibit identical infrared spectra. In the FTIR spectrum, as presented in the figure: 4.10, for the bare ZnO and Mn doped samples shows similar feature for all samples. Pure ZnO and Mn-doped ZnO have wurtzite structure and are further supported by FTIR. Undoped ZnO, Mn doped ZnO (Mn-1, Mn-3, Mn-5) exhibit similar FTIR spectra and the corresponding broadening peaks due to stretching and vibration of different elements present are presented in the table: 4.7.

The identical IR peak for metal-oxide band stretching is observed near 500 cm^{-1}. It has been noted that all samples exhibited IR peak near 500 cm^{-1}, but from the comparative FTIR plot (figure: 4.10) it was clear that compared to other samples, Mn-3 sample exhibited minimum %transmittance at ~500 cm^{-1} due to Zn-O and Mn-O stretching and bending. Observing the intense nature of the IR peak broadening for the sample Mn-3, it can be speculated that Mn concentration plays an important role in shifting IR peak position.

Table 4.7: FTIR peak broadening positions corresponding to different molecular bands for the samples (A) ZnO, (B) Mn-1, (C) mn-3, (D) Mn-5.

Elements	ZnO (cm^{-1})	Mn-1 (cm^{-1})	Mn-3 (cm^{-1})	Mn-5 (cm^{-1})
C-H	832	839	834	763
ZnO	1403	1415	1383	1408
C=O	1554	1473	1494	1630
Co_2	2372	2378	2367	2378
H	2844	2853	2849	2849
C	2914	2919	2919	2929
OH	3413	3418	3385	3429

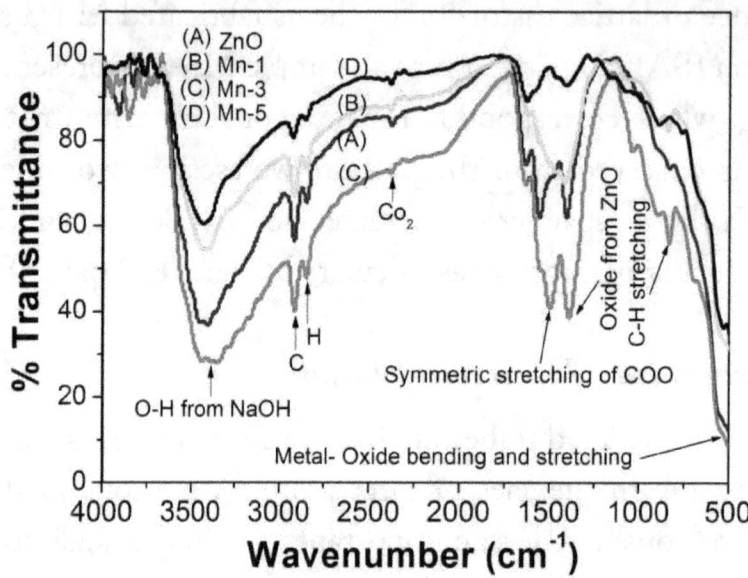

Figure 4.10: FTIR spectrum showing IR peaks for OH, C, H, Co_2, COO, C-H, Oxide from ZnO and metal-oxide stretching and bending corresponding to the samples(A) ZnO, (B) Mn-1, (C) Mn-3 and (D) Mn-5.

IR peak for O-H stretching is obtained within the range 3385 cm^{-1} -3429 cm^{-1}, this 'O-H' band arises due to NaOH which was used as precursor during synthesis. Zinc acetate and Manganese acetate stretching were utilized as reacting components during synthesis, for this symmetric 'COO' stretching IR peaks have been observed within 1473 cm^{-1} -1630 cm^{-1}; The IR peaks due to atmospheric Co_2 are counted within the range 2367 cm^{-1} – 2378 cm^{-1}; for oxide from ZnO related IR peaks are observed within 1383 cm^{-1} – 1415 cm^{-1}; corresponding IR peaks for C-H stretching are seen within 763 cm^{-1} -839 cm^{-1}. From the table: 4.7, it is evident that there is a considerable IR peak shift for all the samples. Shifting of IR peak may contribute some important information regarding the samples. There are two basic thoughts in interpreting the so-called positional fluctuation of peaks of IR spectra under the influence of environmental factors, i.e., temperature and concentration. The peak position change may be caused by the actual frequency shift of a single absorption band or alternatively by the relative intensity changes of overlapped bands. The classical view that has been widely accepted for many decades in the field of vibrational spectroscopy is that gradual changes occur in the vibrational frequency associated with a specific chemical bond. Many previous works concerning the peak position shift of IR spectra are based on the notion that the extent of frequency shift can be directly correlated with the level of specific molecular interactions, such as hydrogen bonding and dipole–dipole interactions [225-231]. The apparent frequency shift of OH stretching or C=O stretching band under temperature or concentration change was attributed to the gradual weakening of such interactions. An alternative view is that the apparent peak position shift is caused by the change in the population of different chemical species. In this case, the position of peak maximum tends to shift due to the variation in the relative intensity contributions of closely overlapped bands with their individual frequencies essentially unchanged.

(iv) Photoluminescence (PL) spectroscopy

'ZnO' have achieved stimulated interests in the last two decades in the field of optoelectronics. Out of many reasons, few of them are:

i.   Possible quantum confinement as observed in nanosized ZnO system may lead to the enhancement in radiative recombination.
ii. Nanostructures of certain morphology with increased surface area and therefore reduced reflection at the air-semiconductor interface, better light extraction may be expected.
iii. From 'ZnO nanostructures' photonic crystal effect may be expected in carefully arranged periodic arrays of nanostructures [232].

The attractiveness of ZnO for optoelectronics lies in its ability to emit light in the ultraviolet (UV) spectral range, which might enable the design of UV light emitting diodes (UV LED's). The recombination processes in ZnO occur via radiative and non-radiative channels. The different types of defects in ZnO may serve as non-radiative recombination centres: point defects, dislocations, surface/interface states etc. Efficient light emission may be obtained via a decrease in the non-radiative contribution. This can be achieved by improved crystal quality of the material. However, even in the case of dominating radiative recombination, the light emission may occur in the UV range (~ 380 nm) as well as in the visible range of spectra (~450 – 650 nm). The latter is so-called defect emission and is commonly observed in ZnO independently of growth technique or substrate temperature [233]. The possible origin of visible emission is deep level defects: oxygen vacancies, zinc interstitials or their combination [234-236]. Despite concerted efforts, the origin of defect emission in ZnO is still under debate.

Photoluminescence (PL) is a powerful technique to explore many properties including light emission properties of semiconductors. In my study the light emission properties of doped and undoped ZnO systems have been studied. The luminescence response as exhibited by doped ZnO is somewhat complicated. This may be due to the presence of various kinds of defects which exist as non radiative centres. The room temperature photoluminescence spectra were studied by using a Perkin Elmer LS-55 spectrometer at excitation wavelength λ_{ex} =325 nm (Xe-source), the data were collected by a computer controlled standard monochromator based photo detector. In some cases we have collected data at different excitation wavelength to compare the result and also to identify Raman peaks which may appear near the λ_{ex}, in the spectrum. PL spectra for undoped ZnO nanords are presented in figure: 4.12. After Gaussian fit for multiple peaks (figure 4.13), in the PL spectra of undoped ZnO specimen, mainly two asymmetrically broadened PL responses were observed: the first, in the UV region at 392 nm and the other in the green region at 425 nm. The UV-band is attributed to the near band edge emission, while the green band is attributed to the emission which has originated due to the intrinsic defect states of ZnO, such as oxygen vacancy, zinc interstitial and oxygen antisite [237,238].

Various assignments of visible emission involve vacancies and interstitial atoms, including cationic and anionic sites in ZnO [239], while some of them reported completely different theoretical and experimental results. At this point, the origin of the defect responsible for the green emission cannot yet be definitely determined only with the experimental results, because visible spectra were greatly affected by preparation methods, environmental conditions, and roles of the coordinating ligands [240, 241]. In general, most defects are considered as a result of the surface states located in the band gap of the nanocrystals, which act as trapping states for the photo generated carriers, while the surface states of the nanoparticles might be investigated by x-ray photoelectron spectroscopy. ZnO has a characteristic green emission peak along with the typical band edge emission peak near UV (E_g = 3.37 eV, λ = 386 nm). On the basis of theoretical and experimental studies P.S. Xu et al. and B.D. Aleksandra et al. [242, 243] have showed different types of defect states related with ZnO nanostructures are:

V_{Zn} = neutral, singly or doubly charged Zn vacancies;

V_o = neutral or singly charged Oxygen vacancies;

Zn_i = neutral or singly charged interstitial Zn;

O_i = interstitial O;

$V_o Zn_i$ = a compex of V_o and Zn_i;

$V_{Zn} Zn_i$ = a complex of V_{Zn} and Zn_i;

O_{Zn} = substitution of O at Zn position;

The singly charged oxygen vacancy (V_o^+) is located at 1.62 eV [24], below the conduction band in the ZnO band gap and results in an emission at ~500 nm. The electron-hole recombination on singly ionized oxygen vacancies is the most widely but not universally accepted mechanism for green luminescence from ZnO. The schematic feature of the energy levels for different defect states in ZnO nanostructure is shown in figure: 4.11. The oxygen vacancies appear to be intrinsic in solution-based synthesis and may result from the abrupt heterogeneous nucleation and growth, mediated by the uneven surface energies. Synthetic protocol and passivation processes are very often responsible for the concentration of surface states. For charge carriers and excitons, these surface states act as surface traps, which generally degrade the optical and electronic properties of the nanocrysystems. For radiative transitions, the surface states can be suitably used. Spanhel, Lubomir and Anderson, Mark A., reported that if the radiative centre is associated in part with surface, their concentration would be expected to decrease with the aggregation of nanostructures [244]. From the PL spectra analysis of my ZnO nanorod systems as presented in figure: 4.12, we speculate that among all the defect states quoted above, neutral or singly charged interstitial Zn (Zn_i) is considered to be the most prominent one which suppress the other defects.

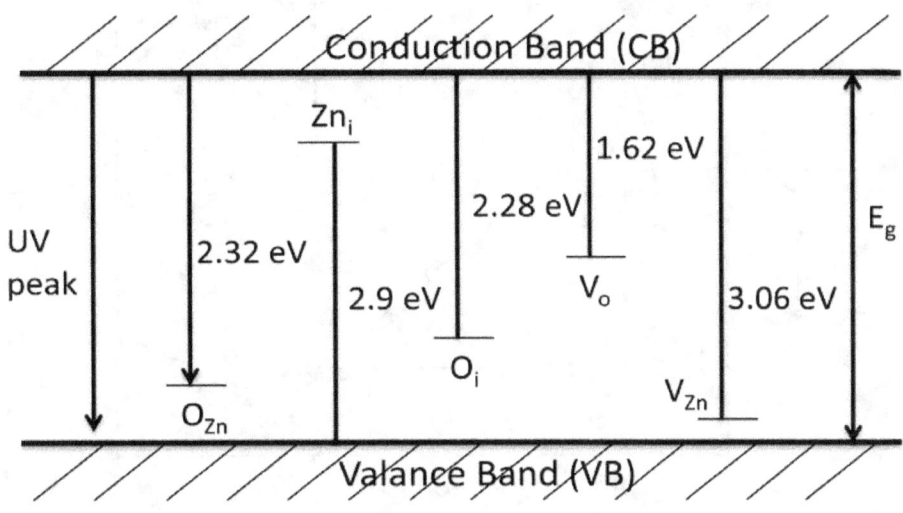

Figure 4.11: The schematic feature of the energy levels for different defect states in ZnO nanostructure.

It is generally accepted that there are two emission bands in the PL spectrum of ZnO. One is in the UV range, which is associated with exciton emission, and another is in the visible range, which originates from the electron–hole recombination at a deep level, caused by oxygen vacancy or zinc interstitial defects [245]. The Gaussian fit for multiple peaks for the PL spectra of bare ZnO is shown in figure: 4.13. The UV peak position and green peak position are recorded at 392 nm and 425 nm respectively. It has been reported that the high-crystallinity and more perfection in surface states may enhance the UV emission in the PL spectrum [246].

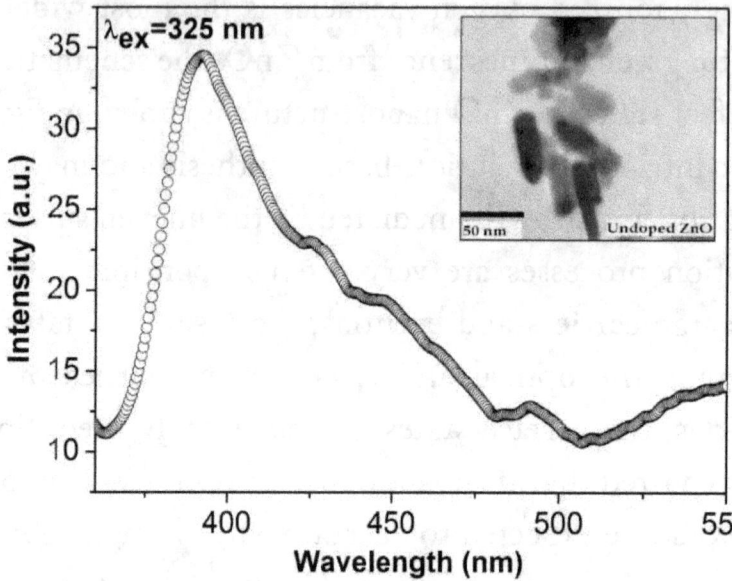

Figure 4.12: Room temperature Photoluminescence (RTPL) spectra of ZnO nanorods with excitation wavelength 325 nm, inset: TEM image of ZnO nanorods.

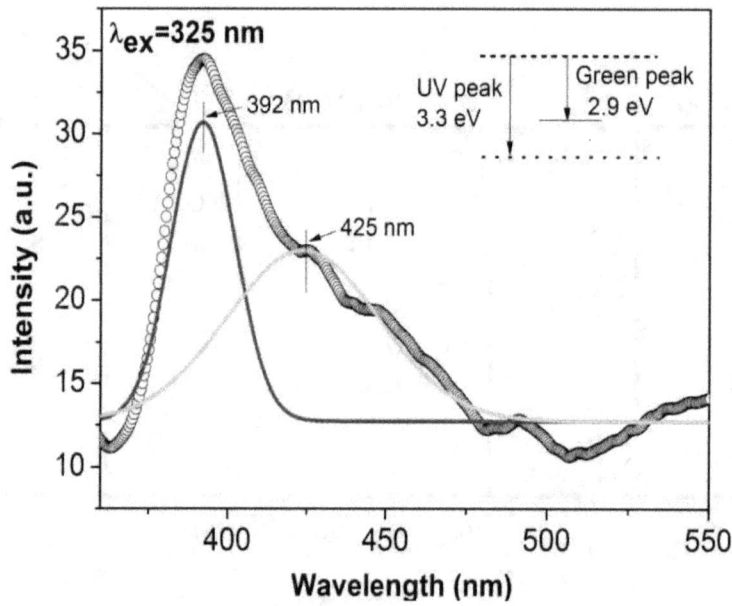

Figure 4.13: RTPL spectra of ZnO nanorods with Gaussian fit for multiple peaks (UV and green) at excitation wavelength 325 nm.

In my case the UV emission is more intense compared to green emission. HRTEM image study also inferred high-cryastallinity nature of the as fabricated Mn-3 sample. The intensity of UV emission is also dependent on the nanostructure size. Below a certain size, the luminescence properties of ZnO nanostructures should be dominated by the properties of the surface [247]. An enhanced UV emission for thinner nanostructures like nanochips has been reported [248], which was attributed to their larger surface area and fewer defects. Thus, fewer defects in the surface and high crystalline nature of the samples respond to the intense UV emission in case of pure ZnO nanostructures.

The optical quality and the possible effects of Mn-doping were investigated using room temperature photoluminescence (PL). A comparison of the PL spectra of the pure ZnO nanorods and that of Mn-doped ZnO nanostructures is illustrated in Figure: 4.14. It has been observed that due to Mn doping the UV peak gets lowered and additional defect related peaks are generated.

In the PL spectra of Mn doped ZnO sample (Mn-3), we have observed three characteristic peaks after Gaussian fit for multiple peaks, which is presented in figure: 4.15. As observed from the plot, the peak positions are recorded as 393 nm, 438 nm and 544 nm. Compared to the PL spectra of bare ZnO, the UV peak position in case of Mn-3 sample remains almost same but the PL intensity is lowered by about three times. The other two peaks are considered to be defect related peaks in the visible region. The peak position at 438 nm may be attributed due to Zn interstitial while at 544 nm due to oxygen vacancy. As noticed from the spectra the peak at 438 nm is more intense while the peak at 544 nm is suppressed. Khalid Mahmood et al.[248] observed a strong UV emission peak around 377 nm for In-doped ZnO nanostructures along with very weak visible emission centred at 530 nm, which indicated the high crystalline quality and low concentration of defects for In-doped ZnO nanostructures.

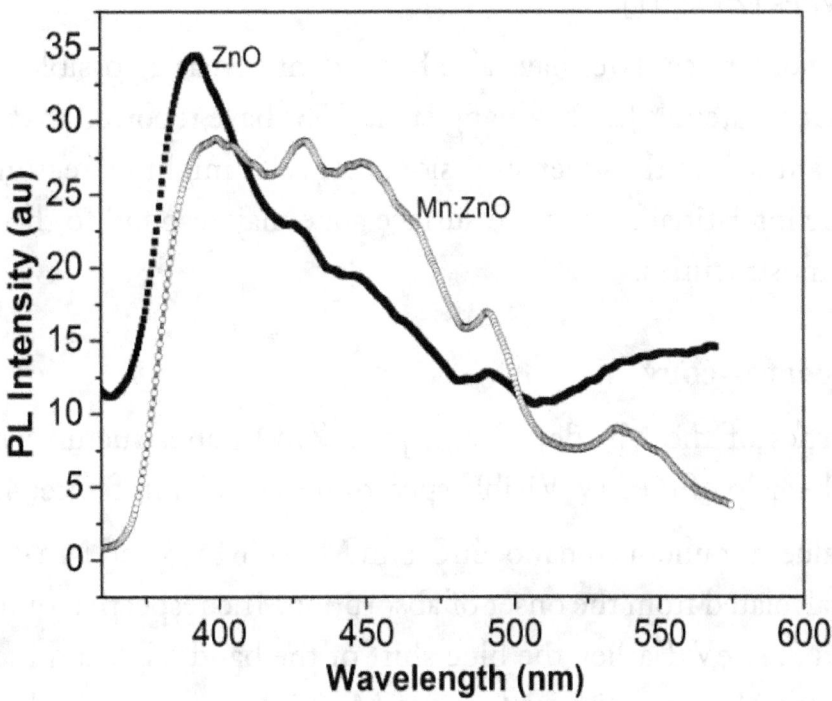

Figure 4.14: RT PL spectra for pure ZnO and Mn:ZnO system at excitation wavelength 325 nm.

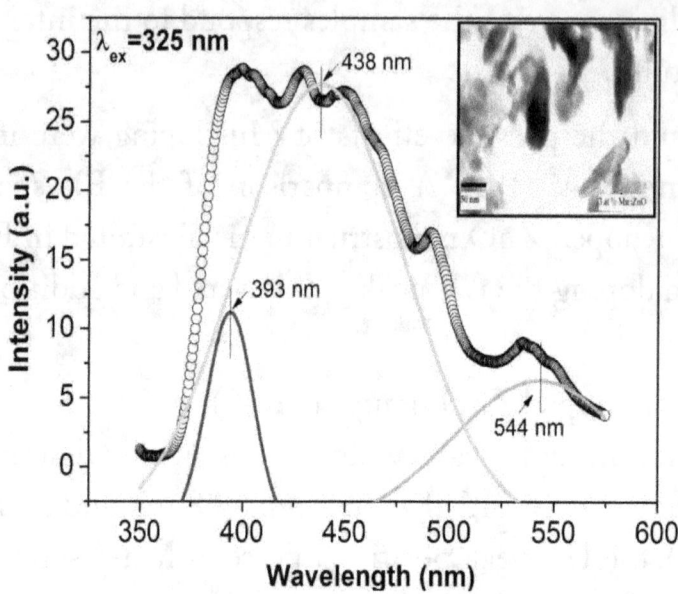

Figure 4.15: Multiple peaks Gaussian fit for RT PL spectra of Mn-3 sample at excitation wavelength 325 nm.

Green emission peak is commonly referred to as a deep level or trap-state emission. The green band is generally explained by the radial recombination of a photo-generated hole with the electron in a singly ionized oxygen vacancy [249]. It has been reported that due to the excess exciton impurity and crystalline defect scattering, there exists a deep-level emission around 2.4 eV in ZnO Nanowires [250,251].

On the other hand, surface states have also been identified as a possible cause of the visible emission in ZnO nanomaterials [252], Zhang et al. [253] have reported that surface states may play a more important role in the green emission. Hence, it might be reasonably inferred that oxygen vacancy, zinc interstitial along with surface state may respond to the green emission of Mn doped ZnO nanostructures.

(v) UV-Visible Spectroscopy

The optical properties of the Mn-doped and pure ZnO nanostructures have been further investigated with the help of the UV-Visible spectrum as shown in figure: 4.16.

The band gap values for undoped nano ZnO and Mn-1, Mn-3 and Mn-5 (1%, 3%, and 5%) doped ZnO were calculated from the onset of absorption. The respective values are found to be 3.52, 3.44, 3.49 and 3.45 eV. Earlier, the blue shift of the band edge and increased hexagonal lattice parameters were shown as the evidence of Mn^{2+} incorporation in the ZnO lattice host [219]. We had not witnessed such effects though a minimal variation in band gap is observed due to Mn-related substitutional doping.

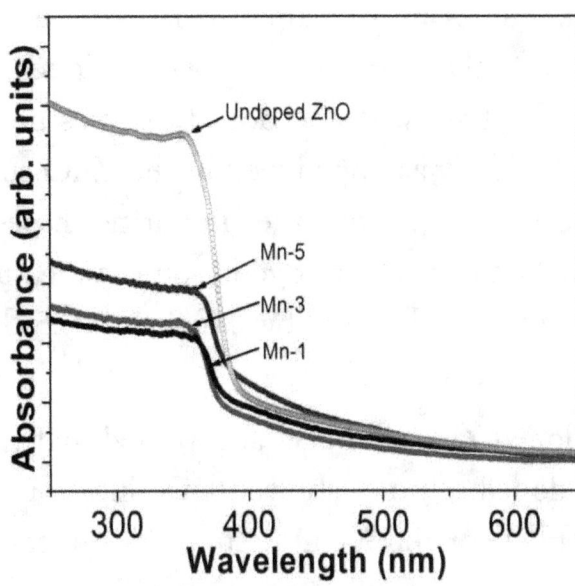

Figure 4.16: UV-Visible spectroscopy of Bare ZnO, Mn-1, Mn-3 and Mn-5 samples.

For wide band gap semiconductors, doping in them often induces dramatic changes in their electrical and optical properties [254, 255] and markedly alters the band gap. According to the theory of semiconductor–metal transition, the band gap energy (E_g) decreases when the impurity is more than the Mott critical density [256]. Hence, heavy doping leads to an obvious narrowing of E_g. It was examined in Figure 4.11 that, with Mn-doping, the UV-visible band shifts to a longer wavelength, which could be possibly due to the narrowing of E_g, which was similar to the results reported by other researchers [257].

4.2 Co doped ZnO

For experimental investigations three different Co doped ZnO samples were prepared by adopting the same solid state chemical reaction route. The Co concentrations in the samples were varied as 1%, 3% and 5% and indexed as Co-1, Co-3 and Co-5.

(vi) X-Ray diffraction studies

The as-synthesized powder samples thus obtained were characterized by powder x-ray diffraction (XRD) by using Philips 1730, Cu-$K\alpha$ radiation, λ =1.54 Å. The X-ray diffraction pattern of undoped ZnO and Co-1, Co-3 and Co-5 samples is presented in figure: 4.17. All the diffraction peaks can be indexed to a hexagonal wurtzite structured ZnO (space group P63mc), without any additional impurity phases, thereby indicating that the wurtzite structure might have not affected due to the substitution of Cobalt. Further, as no excess peaks were detected, it has been concluded that all the starting organic precursors might have been completely decomposed and the Co ions successfully occupy the lattice site rather than interstitial ones. Previous report revealed that XRD study of Co powder alone shows one intense peak at around 44° [258]. In my XRD study with Co doped samples no such peak around 44° has been detected, it indicated that Co was doped well in the ZnO sublattice. However we can't deny the existence of Co, CoO clusters or other impurity phases in the samples since the sensitivity of the X-ray diffractometer may not go beyond to measure their existence. Further, the existence of Co phases had been detected by J. Cui et al.[259] in their Co:ZnO samples annealed above 700⁰C, but in my case, fabrication of all samples were done below 100° C.

In a recent work on Co doped ZnO, B. Pal et al. reported that compared to undoped ZnO, the XRD intensities of Co doped samples showed lowering of intensity and increase in full width at half maxima (FWHM) of the XRD pattern. In my case, we have observed slight decrease in the intensity of doped samples but no clear evidence of increase in FWHM for all Co doped samples has been detected. As noted from the table 4.3, considering the reflection from the plane (101), only Co-1 sample shows maximum FWHM. Further, as observed from the XRD pattern, the sharpness of the peaks reflected from the crystallographic planes (004) and (202) are considerably lowered for Co doped samples.

It has been noticed from a slow scan comparison of the (101) peak of Co-doped and the undoped ZnO NSs that the centres of diffraction peaks of doped ZnOs shift towards high angle compared to undoped ZnO (figure: 4.18). The diffraction peaks along with their relative intensities are found to be in agreement with those from the other reports [260,261]. The shift of peak position is attributed due to the change of size and strain of the NS for incorporation of Co in the ZnO lattice host induced by mechanical stressing during synthesis. Since the ionic radius of Co and Zn are 0.72 Å and 0.74 Å respectively, which are very close, Co doping induced strain is expected to be less significant. However, as a result of ball milling, a compressive strain is introduced in the ZnO NPs [262]. We estimated a reduction in interplanar spacing of ~0.29% from the measured shift in 2θ for (101) plane. This strain in the NPs is expected to influence the electronic, optical, magnetic and other properties including band-structure of ZnCoO.

Table 4.8: Relative data showing FWHM and 2θ corresponding to the plane (101) for determination of crystallite size and interplanner spacing of ZnO, Co-1, Co-3 and Co-5 samples.

	FWHM (in degree)	2θ theta (in degree)	Crystallite size (nm)	d(101) Å
ZnO	0.4985	36.1464	16.5701	2.4823
Co-1	0.5347	36.2499	15.4537	2.4751

| Co-3 | 0.4903 | 36.3543 | 16.8575 | 2.4683 |
| Co-5 | 0.4923 | 36.3459 | 16.7879 | 2.4689 |

Further, the average crystallite size of all the samples was estimated using the most intense diffracted peak (101) broadening technique and are found to be in the range of 15.45 nm to 16.86 nm. Table: 4.8 shows the FWHM and corresponding 2θ values corresponding to the plane (101) for calculation of crystallite size of as synthesized nanostructures (NS). The sizes estimated from the XRD data for bare ZnO, Co-1, Co-3 and Co-5 samples are 16.57 nm, 15.45 nm, 16.86 nm and 16.79 nm respectively.

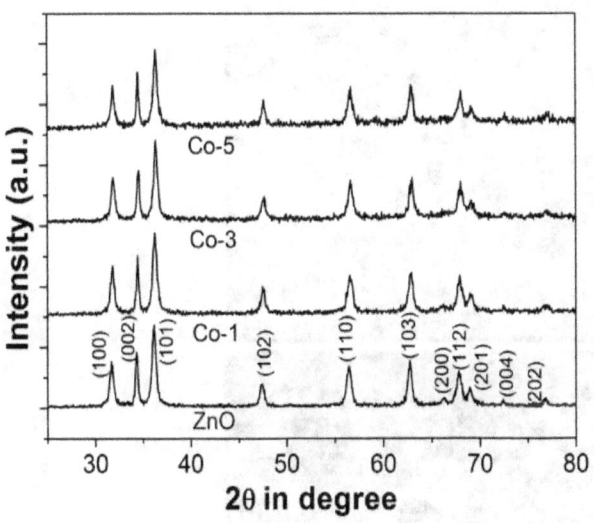

Figure 4.17: XRD pattern of Co-1, Co-3, Co-5 and bare ZnO samples.

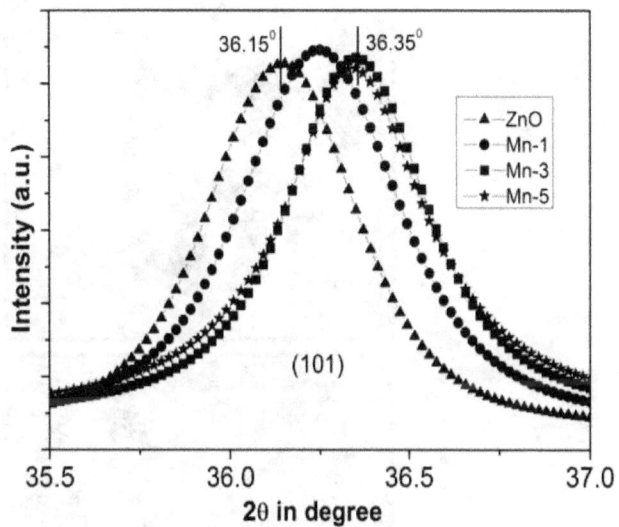

Figure 4.18: XRD pattern showing shifting of the centres of (101) diffraction peaks.

(vii) Electron Microscopy study

(a)   Co-1

Structural study of the bare ZnO and Co doped ZnO was further extended by performing Transmission electron microscopy (TEM) study. A clear visual evidence of the formation nonords with the bare ZnO and Co doped ZnO is being depicted by electron microscopy study. The TEM images of the as fabricated Co-1 sample are presented in figure: 4.19. At some lower magnification formation of rod like nanostructures were found [figure: 4.19(A)], while at higher magnification on a selected section, few irregularly shaped rods were confirmed (figure: 4.19B). Using standard software the length and diameters of few rods were measured from which we found that the average length and diameter of the nanorods so formed lie within range 80-100 nm and 10-30 nm respectively. By adopting Ball milling method other researchers found different shaped nanoparticles for Co doped ZnO cases, particularly for bare ZnO they found mostly spherical nanoparticles [258,261]. Since, we have adopted mechanical milling (manual) and interest was paid in unidirectional motion of milling during synthesis of Co:ZnO system. Probably this may be the reason why we are getting elongated nanostructures of my samples. The corresponding HRTEM image as shown in figure: 4.19C indicates the lattice space 2.4 Å, matching the space of the lattice plane (101) for the wurtzite ZnO. It confirms that the preferable crystal growth direction is (101). The HRTEM image reveals existence of defect states in the surface of nanords as fabricated. Further, no lattice distortion was observed in the SAED pattern

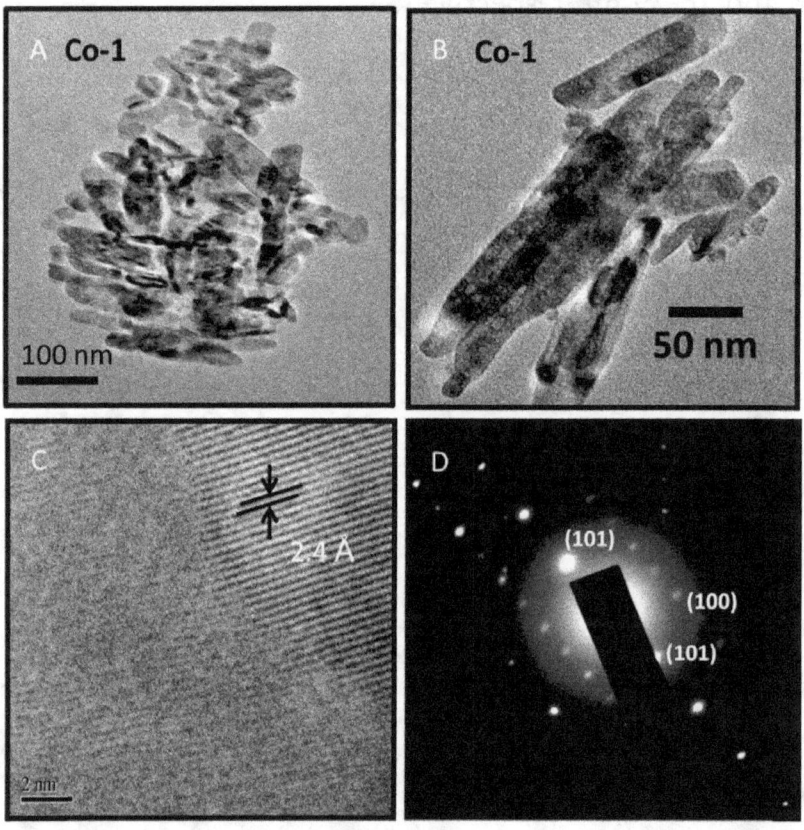

Figure 4.19: (A) TEM image of Co-1 sample (A) rod like nanostructures at lower magnification (B) Few Co-ZnO (1at% Co) Nanorods at higher magnification (C) High resolution TEM image of a small crystallite showing surface defect without lattice distortion with lattice spacing ~2.4 Å corresponding to (101) plane (B) Electron diffraction pattern (SAED) obtained from the same section of the Co-1 sample.

The selected area electron diffraction pattern (SAED) for Co-1 is presented in Figure: 4.18(D). All values for d-spacing calculated from SAED pattern are close to ZnO structure, indicating wurtzite phase of the sample. It clearly indicates non-existence of Co, CoO clusters or other impurity phases in selected area of the nanorod. As the SAED pattern was obtained by focussing the beam on a few nanostructures of the sample and also due to law concentration of Co doping, there was no Co clusters observed in the as prepared nanostructures.

(b) Co-3

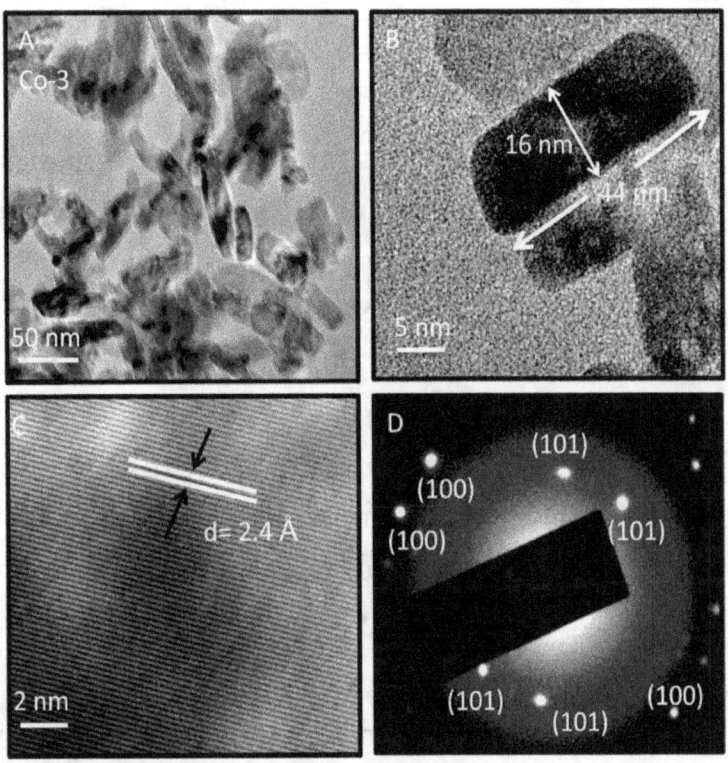

Figure 4.20: (A) TEM micrograph of Co-3 sample, (B) An isolated nanorod obtained from Co-3 sample with length ~44 nm and diameter ~16 nm, (C) HRTEM imageof Co-3: Beam focussed on the small crystallite showing clear lattice fringes without any lattice distortion with lattice spacing ~2.4 Å corresponding to (101) plane, (D) Electron diffraction pattern obtained from the same section of the Co-3 sample.

The high resolution TEM images of Co-3 sample are presented in figure: 4.20. It could be seen from the micrograph of figure: 4.20A that the nanoclusters are generally elongated with average length 71 nm and diameter 19 nm. During the study many isolated rods for Co-3 sample were detected. As indicated in figure: 4.20B, high resolution of TEM image focussed on a single rod with length 44 nm and diameter 16 nm has been observed. It has been noticed that compared to Co-1 sample, elongated shapes are more regular in case of Co-3 sample. The high resolution TEM of the small crystallite is presented in figure 4.20C, it indicates uniform lattice structure with clear lattice spacing ~2.4 Å corresponding to the plane (101). Thus, we can infer that the growth direction of small crystallite is along (101). Further, from the formation of clear lattice fringes, we can conclude that there is no lattice defect and the nanoclusters are single crystallite state. The estimated lattice spacing 'd's are a little bit higher than those of bulk ZnO, suggesting that the Co atoms are substitutes. Several separated rings in the electron diffraction pattern from the inset are identified as being consistent with wurtzite ZnO structure, implying the unchanged structure of Co-doped ZnO clusters to that of wurtzite ZnO structure [263].

(c) Co-5

Structural investigation of Co:ZnO system was further extended higher concentration of Co. The TEM micrographs and HRTEM image and SAED pattern for 5 at% Co doped (Co-5) is presented in figure: 4.21. For this sample, though nanostructures are irregularly shaped and agglomerated, most of the nanostructures are identified to be elongated in shape (figure: 4.21A). Few isolated well shaped nanorods were found during TEM investigation. One such isolated rod with length ~70 nm and diameter ~16 nm is presented in figure: 4.21B. The corresponding HRTEM image as shown in figure: 4.21C indicates the lattice space 0.24 nm, matching the space of the lattice (101) for the wurtzite ZnO at diffraction angle, $2\theta = 36.68^0$ in the XRD pattern. Single crystalline nature of the as prepared sample (Co-5) is indicated from its HRTEM image.

Fine structure and clear fringe pattern as observed in this image depicts that the surface of the nanocrystal is free from any defect. Few diffracting points as noticed from the SAED pattern (figure4.21D) of the sample are indexed for (101) and (100) planes.

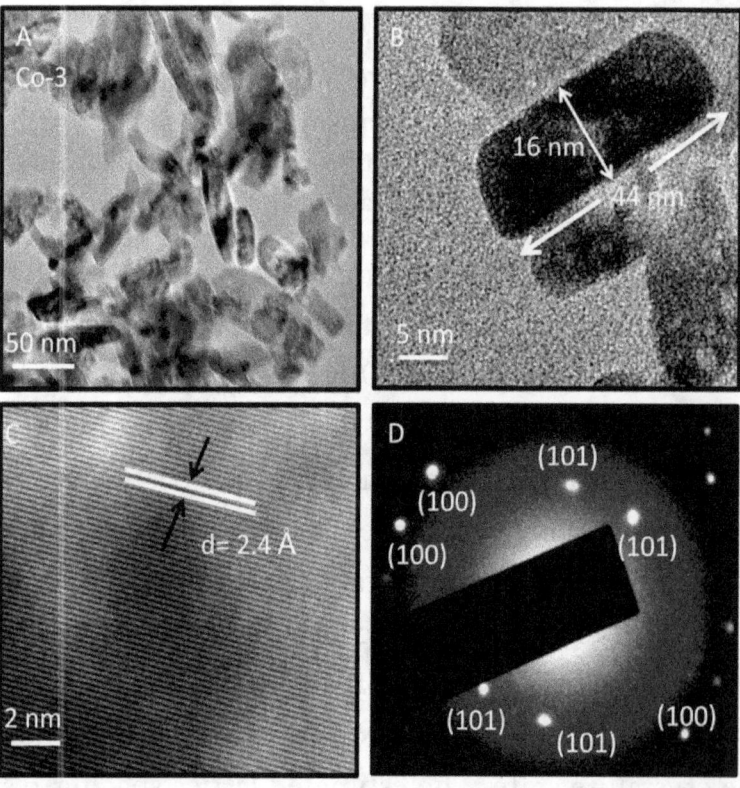

Figure 4.21: (A) TEM micrograph of Co-5 sample, (B) An isolated nanorod with length ~70 nm and diameter ~16 nm. (C) High resolution TEM image: Beam focussed on the small crystallite showing no lattice defect with lattice spacing ~2.4 Å corresponding to (101) plane (D) Electron diffraction pattern obtained from the same section of the Co-5 sample.

(viii) Energy dispersive X-ray study (EDX)

Energy dispersive X-ray study (EDX) gives the signature of atoms of different materials contained in the sample along with their relative content. The EDX analyser produces a spectrum of the elements present in targeted areas of the samples allowing detectable elements to be quantified or mapped. We have performed EDX study for the samples ZnO, Co-1, Co-3 and Co-5 and is presented in figure: 4.22. During the spectrum processing, no extra peaks were omitted and all elements were analyzed.

In general EDX is studied as an additional option with SEM study. For EDX investigation, my materials were coated with carbon, for which carbon 'C' related peaks were observed in each spectrum.

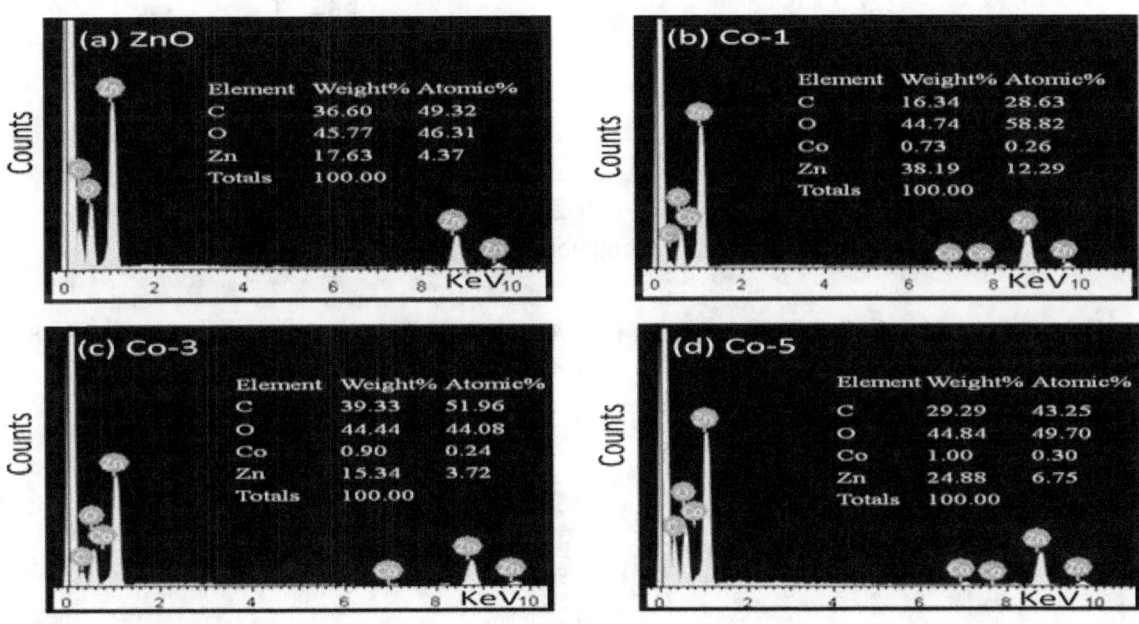

Figure 4.22: Energy dispersive X-ray study (EDX) of the samples (a) ZnO, showing Zn and O related peaks and corresponding amount detected in few nanostructures.(b) Co-1, (c) Co-3 and (d) Co-5: showing Zn, O and Co related peaks with their relative amount.

Figure: 4.22(a) represents EDX spectrum for pure ZnO sample, where Zn and O related peaks gives the evidence of purity of the sample. On the other hand, in the spectrum as presented from Figure: 4.22(b) to 4.22(d), we have noticed the presence of Zn, O and Co related peaks. It is to be noted that EDX investigation was performed with only few particles of each sample, which indicated that cobalt ion was distributed in the entire sample. One can note from the tables cited along with each spectrum that the amount (wt %) of Co is increasing linearly from Co-1 to Co-5 sample. Moreover, a slight deviation from the fixed amount as given during synthesis is observed during the study. This may be due to the reasons, (a) Only a few particles were exposed to the beam, (b) We cannot neglect the instrumental error, (c) Agglomeration may occur due to the period between synthesis and characterization.

(ix) Fourier Transform Infrared spectroscopy (FTIR) study

In the FTIR spectrum, as presented in the figure: 4.23, for the bare ZnO and Co doped samples shows similar feature for all samples. Pure ZnO and Co-doped ZnO have wurtzite structure and are further supported by FTIR.

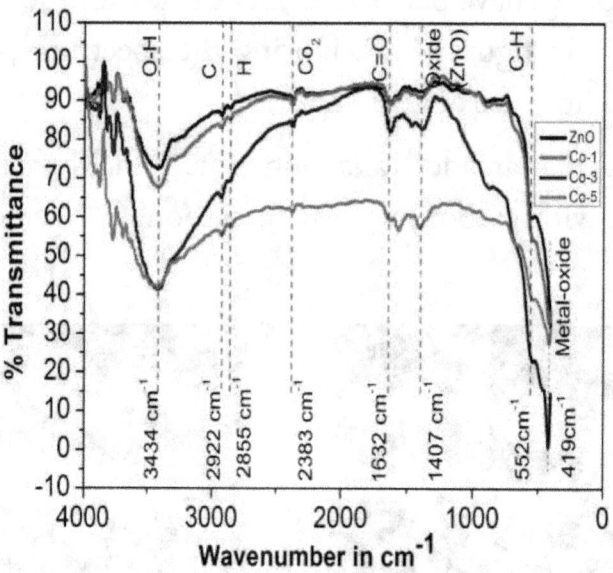

Figure 4.23: Individual FTIR spectrum for (A) ZnO: (B) Co-1, (c) Co-3 and (D) Co-5.

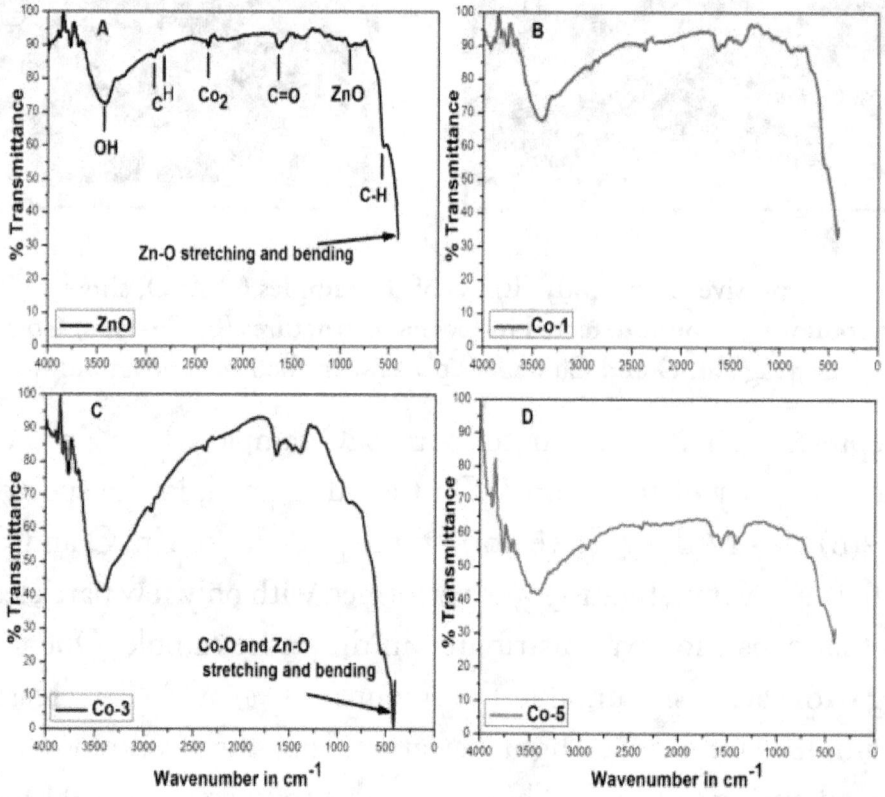

Figure 4.24: Comparative FTIR spectrum for (A) ZnO: (B) Co-1, (c) Co-3 and (D) Co-5.

Undoped ZnO, Co doped, similar spectra and the corresponding broadening peaks at 3434 cm^{-1} is due to OH Stretching vibrations (from NaOH, used as a precursor), the band at absorption of atmospheric CO_2 on the metallic cations at 2383 cm^{-1}; the band at 1632 cm^{-1} represents to C = O stretching vibrations; the band at 1407 cm^{-1} indicates oxide from ZnO peak and the band observed at 955 cm^{-1} is attributed to C-H band [264, 265]. It is well known that for metal-oxide band stretching the identical absorption peak should present near 500 cm^{-1}. In my case we have observed this identical absorption peak for all samples at 419 cm^{-1}.

In figure: 4.24, a comparative FTIR plot is presented where IR peak positions for all samples (ZnO, Co-1, Co-3, Co-5) corresponding to O-H, C, H, Co^2, ZnO, C=O, C-H and metal-oxide stretching and bending are shown with respective dash lines. In addition to this, from the FTIR spectrum (figure: 4.22 and figure: 4.23) it has been observed that Co-3 sample exhibit zero %transmittance at 419 cm^{-1} due to Zn-O and Co-O stretching and bending. From this, it can be concluded that Co-3 sample behave differently compared to any other sample. Moreover, the ZnO sample shows slight deviation (~ 57 cm^{-1}) for the IR peak corresponding to C=O stretching. Except this slight deviation no other peak shift has been observed. No shift of IR peaks indicates that the FTIR result is in good agreement with that of XRD [266].

(x) Luminescence spectroscopy

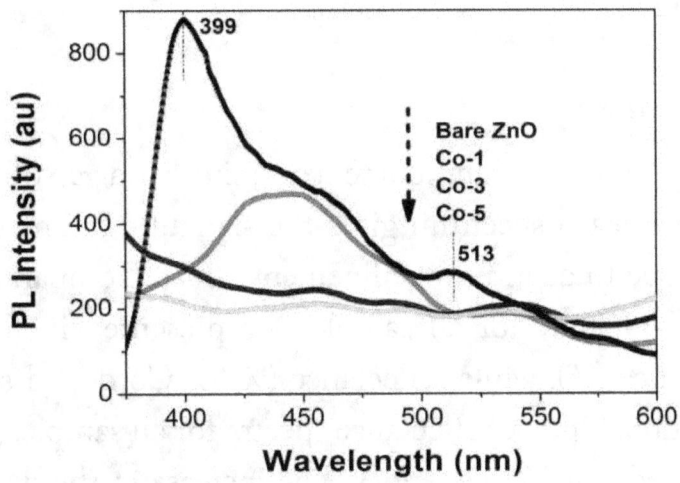

Figure 4.25: Photoluminescence spectra of different samples: ZnO, Co-1, Co-3 and Co-5.

Photo luminescent properties of bare ZnO and Co doped (Co-1to Co-5) ZnO is presented in figure: 4.25. Photoluminescence investigation, performed on the bare ZnO specimen, at an excitation wavelength of ~325 nm showed two main PL bands: one in the UV region

(~399 nm) and the other broad peak in the green region (~513 nm). An associated lowering of intensity of the doped samples indicates suppression of radiative recombination process.

For Co-1 and Co-3 doped samples red shifting is associated with suppression of the UV and green band emission peaks. For Co-1 doped ZnO nanorods, suppressed but broadened UV emission peak centered at around 440 nm were observed.

The UV peak is red shifted and suppressed about 46% as compared with the bare ZnO sample, for the Co-3 doped sample the suppression decreases by about 48% and the peak width was broadened further. However, when Co-5 was doped no characteristic peak for near band edge emission was observed. Similar feature has been observed by Y. X. Wang et al. [267].

In general, the intensity ratio of UV emission band to visible emission band is considered to be as an indicator of the crystallinity of ZnO materials [268]. For better crystallinity this ratio should be higher. In a similar work done by Z. W. Zhao and B. K. Taya et al. [269] have found decreased ratio of UV emission band to visible emission band which indicated inferior quality of corresponding sample. In my study the Co doped ZnO sample exhibited relatively lower peaks in the visible range while relatively higher peaks at UV region. This leads to increased ratio of UV emission to visible band emission, which in turn suggests good crystallinity of my samples. Generally higher order peak in the visible region arises due to defect states associated with oxygen vacancies or Zn interstitials or other defects induced by the dopant or dopant-related defects in the samples which further support the higher intensity in absorbance [270].

(xi) UV-Visible Spectroscopy

UV- Visible spectroscopy is an important tool to determine band gap of as synthesized samples. The onset of absorption of each spectrum gives the signature of enhancement of individual band gap of the sample due to quantum confinement effect. Comparative reflectance spectra (%) obtained from UV-Vis study for all samples are presented in figure: 4.26. Bare ZnO indicates highest reflectance (%), while reflectance (%) of Co doped samples decreases with increasing Co concentration. Optical reflectance spectra for all samples reveal strong quantum confinement with increased band gap energies. The increase in the diffuse reflectivity with a definite linear region of the greatest slope is attributed to an exponential drop in the absorption coefficient. The onset of this exponential drop had been suggested as a more universal method of determining absorption edges from which the band gap can be deduced. For calculation of absorption edge a linear fit was given to the linear part of the increase in reflectivity [271].

Band gap of all Co doped and bare ZnO were calculated from the intersection of two linear portions of each spectrum which showed blue shift. For bare ZnO (figure: 4.27a) the enhancement is small (~0.02 ev), the enhancement energies estimated for (Co-1 to Co-5) were 0.05 eV, 0.04 eV and 0.06 eV respectively (Figure:4.27b to 4.27d). The band gap of all Co doped samples exceeds than pure ZnO, this feature agrees with the value reported earlier [272-275].

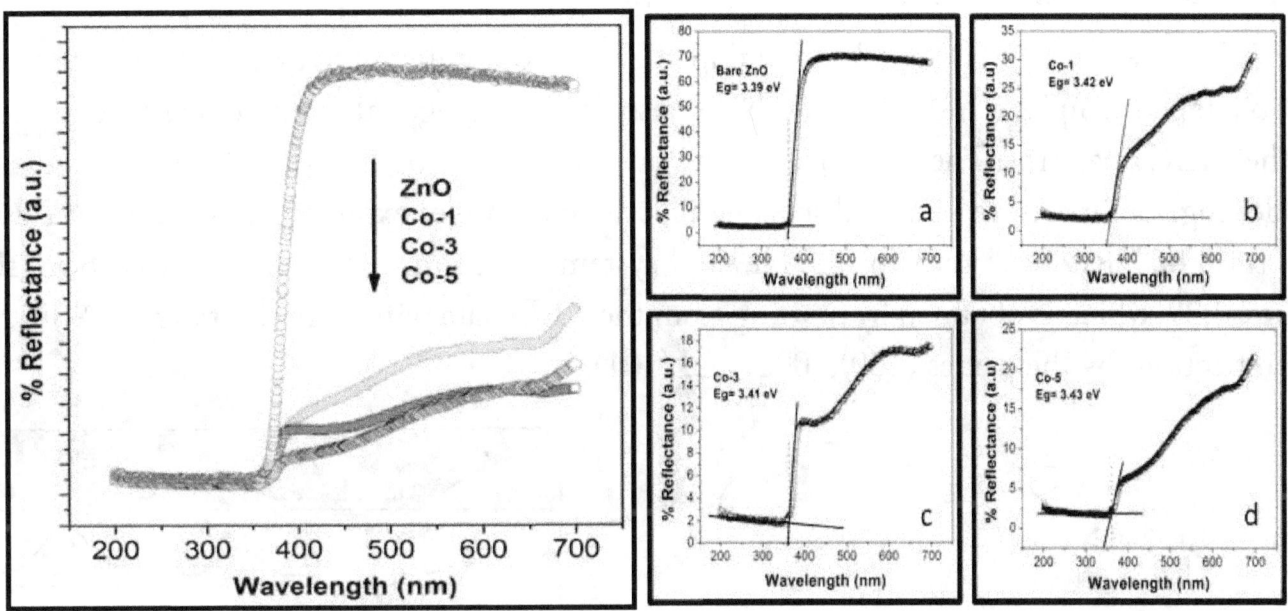

Figure 4.26: Comparative UV-Vis spectra showing % Reflectance of different samples: ZnO, Co-1, Co-3 and Co-5.

Figure 4.27: the optical reflectance spectra for the samples (a) ZnO, (b) Co-1, (c) Co-3 and (d) Co-5.

Enhancement of energy due to quantum confinement of the nanostructures has been indicated in the UV-Visible study.

4.3 Ni doped ZnO

Among the transition metal, Ni is an important dopant; since, Ni $^{2+}$ (0.69 A°) has the same valence as Zn^{2+}and its radius is close to that of Zn+ (0.74 A°), so it is possible for Ni^{2+} to replace Zn^{2+} in ZnO lattice. So far, the influence of Ni doping in the ZnO lattice host has been reported by several groups. Doping of Ni in ZnO matrix has been done by several techniques [276,277], we have fabricated Ni doped ZnO nanpstructures by adopting the same solid state chemical reaction route as described in chapter-2. The concentration of Nickel acetate and Zinc acetate were varied within the range 1 at% Ni - 5 at% Ni and accordingly indexed as Ni-1, N-3 and Ni-5. In this chapter the structural and optical properties will be focussed.

XRD. EDS. HRTEM were carried out to investigate structural properties, while PL and UV-Vis. Spectroscopy study were performed to investigate optical properties of Ni:ZnO nanostructures. XRD study ruled out the existence of additional phases in the sample due to Ni doping. HRTEM study showed the clear evidence of formation nanorods for Ni:ZnO system.

(xii) X-Ray diffraction study

The XRD spectrum of Ni doped and undoped ZnO is presented in figure: 4.28. Like Mn and Co, Ni doped samples (Ni-1, Ni-3 and Ni-5) also exhibit hexagonal wurtzite structure of ZnO in their XRD patterns. The XRD pattern reveals the absence of any secondary phases due to Ni doping. Compared to undoped specimen Ni doped samples exhibit peak shift in the XRD spectra. The XRD peak shift for the Ni:ZnO system can be more clearly visualized from the figure: 4.29, which is obtained by plotting from the XRD data within the '2θ' range 31°-37⁰ for the reflections by the planes (100), (002) and (101).

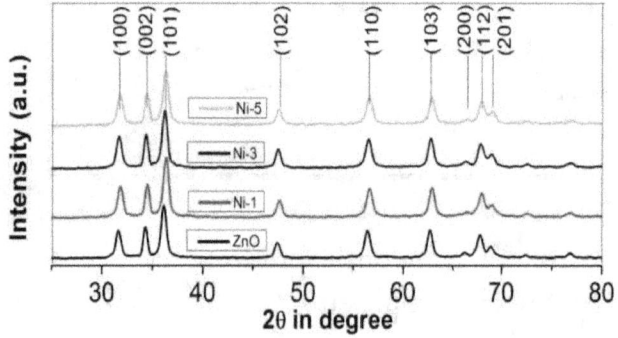

Figure 4.28: XRD pattern for the samples (ZnO, Ni-1, Ni-3 and Ni-5)

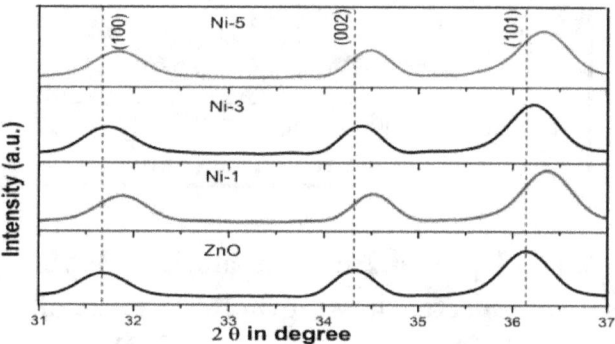

Figure 4.29: Enlarged view of the XRD pattern for the samples within the range 31° - 37° for planes (100), (002) and (101).

Table 4.9: '2θ' values corresponding to the peak positions for diffracting planes (1 0 0), (0 0 2) and (1 0 1) for the samples (ZnO, Ni-1, Ni-3 and Ni-5)

Plane	ZnO	Mn-1	Mn-3	Mn-5
(1 0 0)	31.68	31.87	31.72	31.83
(0 0 2)	34.32	34.5	34.39	34.48
(1 0 1)	36.14	36.36	36.22	36.33

The values for '2θ' at the peak positions corresponding to the diffracting planes (1 0 0), (0 0 2) and (1 0 1) for the samples (ZnO, Ni-1, Ni-3 and Ni-5) are presented in the table: 4.9. Different values for '2θ' corresponding to a particular peak clearly indicates the peak shift.

The shift of peaks for the Ni doped samples take place towards the higher value of '2θ' from the peak positions for pure ZnO sample. The change of XRD peak can be attributed for the change of size and strain of the Ni:ZnO nanostructures due to Ni doping.

Table 4.10(a): Lattice spacing for the diffracting planes (1 0 0), (0 0 2) and (1 0 1) for the samples (ZnO, Ni-1, Ni-3 and Ni-5) estimated from their respective '2θ' values as shown in table: 4.9.

('d' spacing in Å)				
	ZnO	Ni-1	Ni-3	Ni-5
(1 0 0)	2.82	2.81	2.82	2.8
(0 0 2)	2.61	2.6	2.6	2.6
(1 0 1)	2.48	2.47	2.48	2.47

Table 4.10(b): Estimated crystallite size of as synthesized nanostructures for the samples (ZnO, Ni-1, Ni-3 and Ni-5) estimated from FWHM and 2θ values corresponding to (101) diffracting plane of each sample.

	FWHM (w) (degree)	2θ (degree)	crystallite size (nm)
ZnO	.53254	36.14	15.51
Ni-1	.55806	36.36	14.81
Ni-3	.54001	36.22	15.3
Ni-5	.5863	36.33	14.1

The lattice spacing 'd' for all samples were estimated from their respective '2θ' values corresponding to the diffracting planes (1 0 0), (0 0 2) and (1 0 1) of the nanocrystals which are shown in the table: 4.10(a). From the table 'd' values were utilizes to estimate the lattice parameters 'a' and 'c' as 3.237 Å and 5.2 Å respectively according to the equation 4.1. For determination of size of the nanostructures, FWHM and 2θ values corresponding to the most intense peak (1 0 1) were considered for the Scherer formula (equation: 3.2) and the respective values are presented in the table: 4.10(b). From the estimated size, we have observed that the crystallite size of the Ni doped ZnO slightly decreases due to Ni doping compared to undoped specimen. The size of the nanocrystals lies within the range 14.1 nm – 15.51 nm.

The size and strain of the nanostructures were also estimated from Williamson-Hall (W-H) plot as shown in figure: 4.30 for the sample Ni-5. In this plot we have considered first six peaks of the XRD pattern. From the intercept (~0.0099) and the slope

($\sim 7.488 \times 10^{-4}$) of the plot the average size and strain of the Ni-5 nanostructures were estimated. In a similar way the size and strain of other samples were also estimated and presented in the table 4.11.

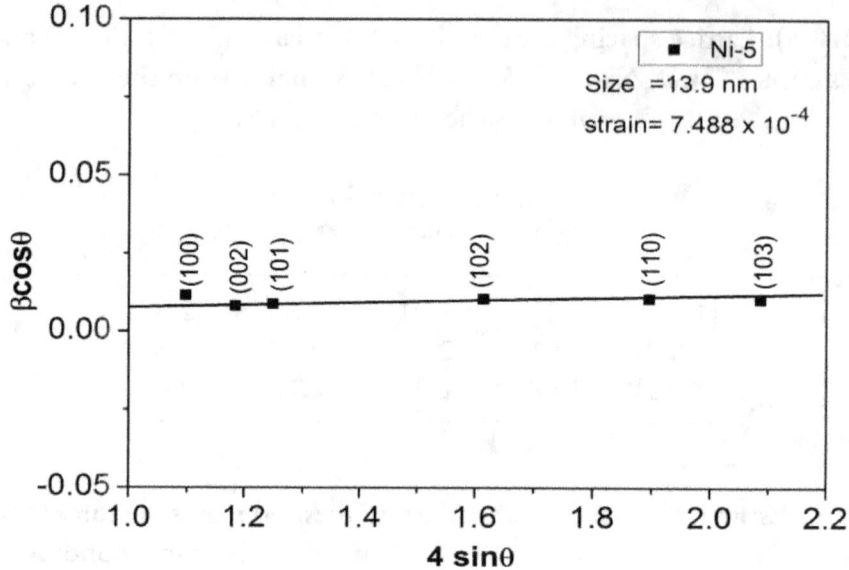

Figure 4.30: Williamson- Hall plot for the sample Ni-5, from the slope and intercept of the straight line strain (7.488×10^{-4}) and size (13.9 nm) of the Ni-5 nanostructures were estimated.

Table 4.11: Crystallite size and strain estimated from Williamson- Hall plot for the samples (ZnO, Ni-1, Ni-3 and Ni-5)

sample	size (nm)	strain
ZnO	16.17	2.58×10^{-3}
Ni-1	15.75	1.14×10^{-3}
Ni-3	15.9	1.44×10^{-3}
Ni-5	13.9	0.75×10^{-3}

From the above investigations it has been noticed that with decreasing size of the nanostructures the strain also decreases. The highest stain was exhibited by ZnO nanorod with size 16.17 nm, while the lowest strain was shown by Ni-5 with size 13.9 nm. The particle size, as obtained from both XRD data is in agreement with the sizes as calculated from W-H plot. This may be due to the limitations of Scherrer formula where particle strain is not considered.

(xiii) Photoluminescence property

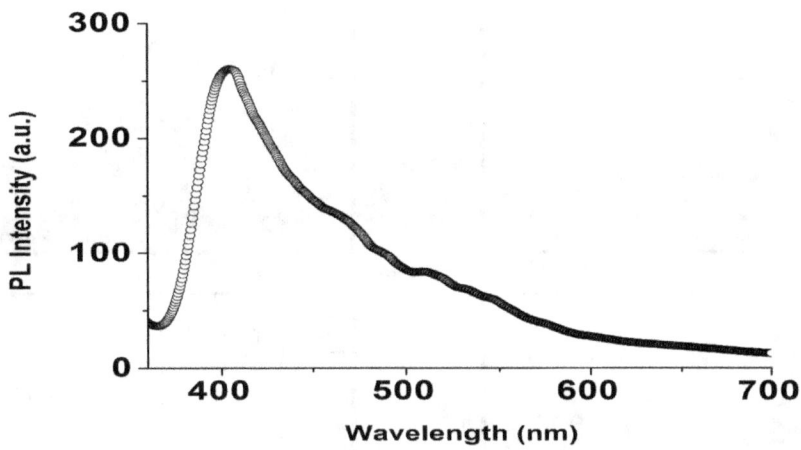

Figure 4.31: Luminescence response of Ni-3 nanorods and nanoparticles.

Few studies on Ni doped ZnO have been reported and several results showed that the luminescence properties of ZnO were changed after doping of Ni [278–281]. Photoluminescence response of Ni-3 sample is shown in figure: 4.31 The intense peak at 404 nm (~3.04 eV) is attributed due to the typical band edge (near ultra violet) emission peak, in addition to this other suppressed peaks are observed at 463 nm (~2.68 eV) and 511 nm (~2.4 eV), which are contributed due to defect states for e.g. the green peak (~511 nm)is due to substitution of O at Zn position [282,243]. The suppressed emission peak in the green region may be attributed due to less defect in the crystal structure of the Ni:ZnO system.

Figure: 4.32 shows the individual optical transmittance spectra for (A) ZnO, (B) Ni-1, (C) Ni-3 and (D) Ni-5 nanostructures. The observed transmittance (%) for the samples was recorded in between 58% to 73% in the UV region. The sample Ni-3 exhibited highest transmittance (73%), while the Ni-1 sample exhibited the lowest transmittance (58%). The transmittance (%) for all samples in the visible region was recorded below 2.5%. The energy gaps (E_g) for the nanorods has been estimated from the spectra using the relation for a direct transition [283], which are 3.453 eV, 3.425 eV, 3.429 eV and 3.426 eV for the samples ZnO, Ni-1, Ni-3 and ni-5 respectively. The comparative transmittance (%) spectra for all samples can be visualized from the figure: 4.33

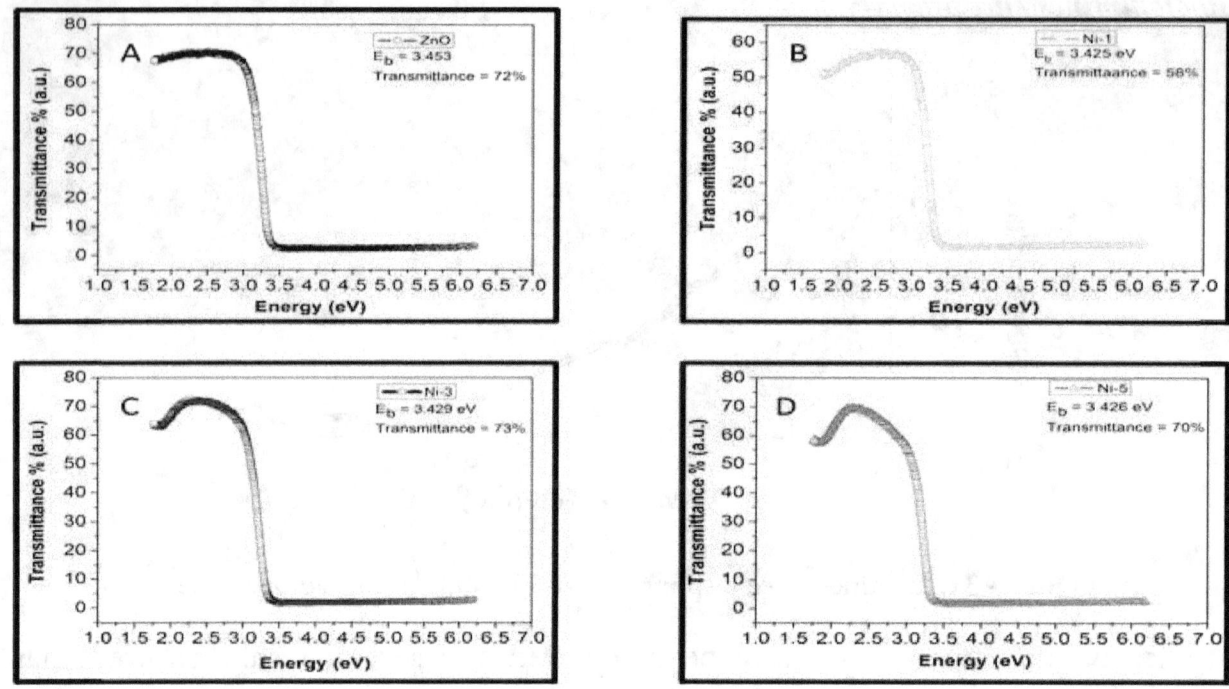

Figure 4.32: % Individual transmitted spectra for (A) ZnO, (B) Ni-1, (C) Ni-3, and (D) Ni-5.

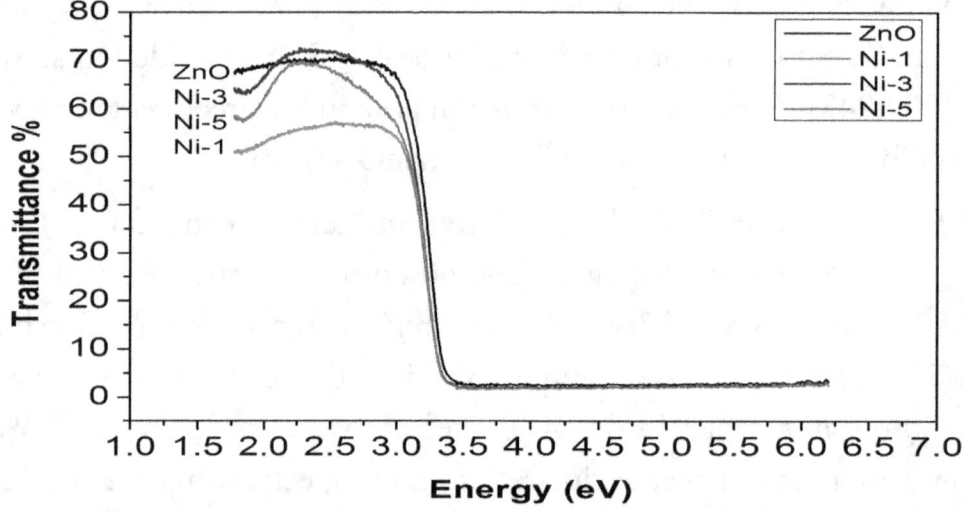

Figure 4.33: % Compared transmitted (%) spectra for (A) ZnO, (B) Ni-1, (C) Ni-3, and (D) Ni-5. The sample Ni-3 shows highest transmittance (%), while the sample Ni-1 shows minimum transmittance (%).

In the %reflectance spectra, it has been observed that due to the strong quantum confinement effect all bare and Ni doped samples exhibited blue shift. There is a slight but noticeable variation in band gap energies among the samples. This slight variation is attributed due to their similar (almost) crystalline sizes.

(xiv) Transmission electron microscopy study

(a) Ni-1

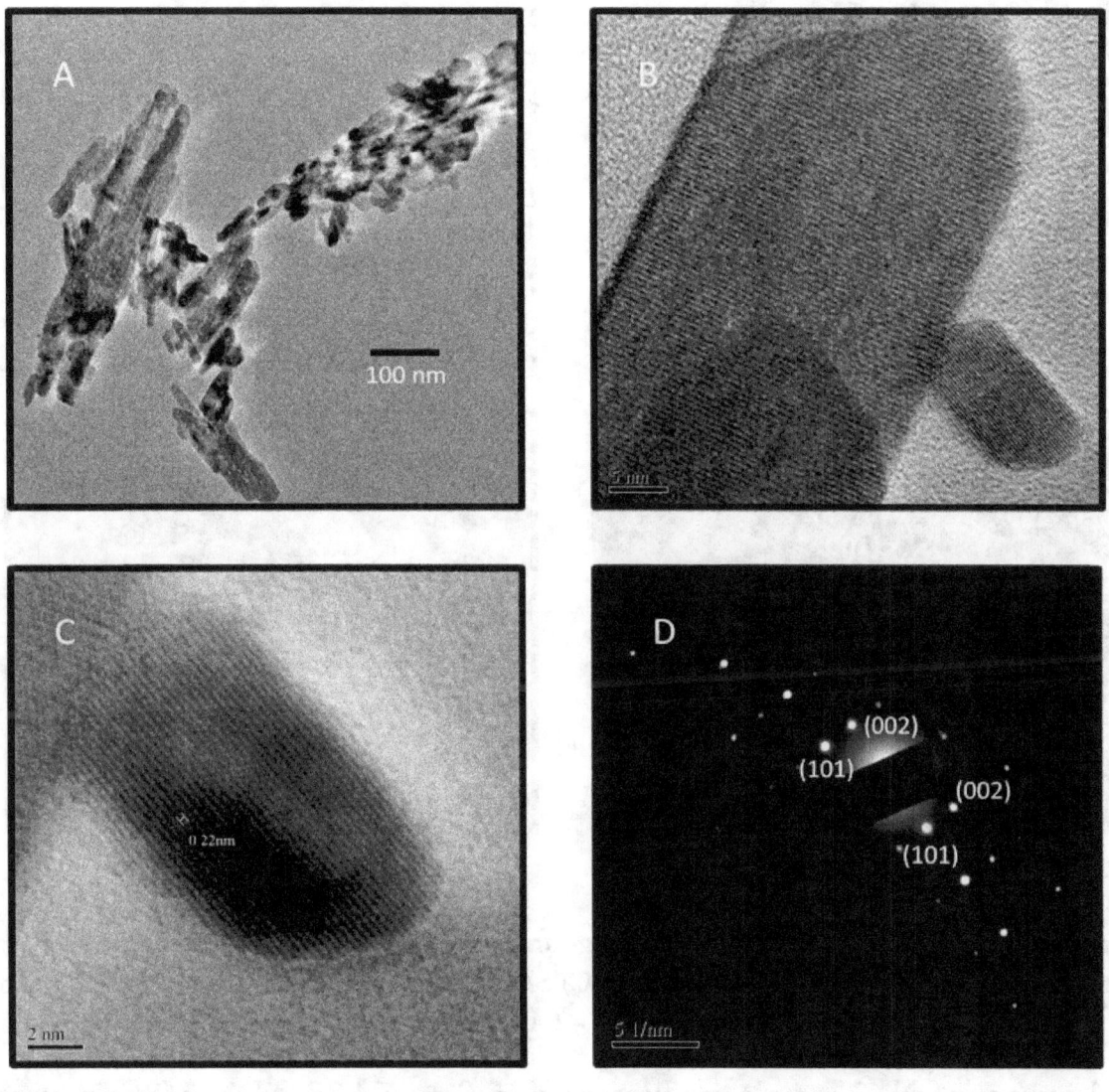

Figure 4.34: (A) TEM image of Ni-1 showing formation of elongated nanostructures, (B) HRTEM image of two isolated nanorods, (C) High resolution image of an isolated nanorod with lattice spacing 2.2 Å. (D) SAED pattern focussed on the nanorod shown in (C).

Formation of elongated nanostructures has been depicted by the TEM study which is presented in the figure: 4.34(A). TEM study confirmed the formation of nanorods with average length 60 nm and diameter 15 nm. High resolution TEM image of an isolated nanorod is shown in figure: 4.34(B) and 4.34(C). Clear lattice fringe indicates good crystalline nature of the sample. The lattice spacing ~2.2 Å corresponds to the crystal growth plane (102). SAED pattern (figure: 4.34D) shows uniform distribution of lattice points. Two inner rings corresponds to the diffracting planes (101) and (002).

(b) Ni-3

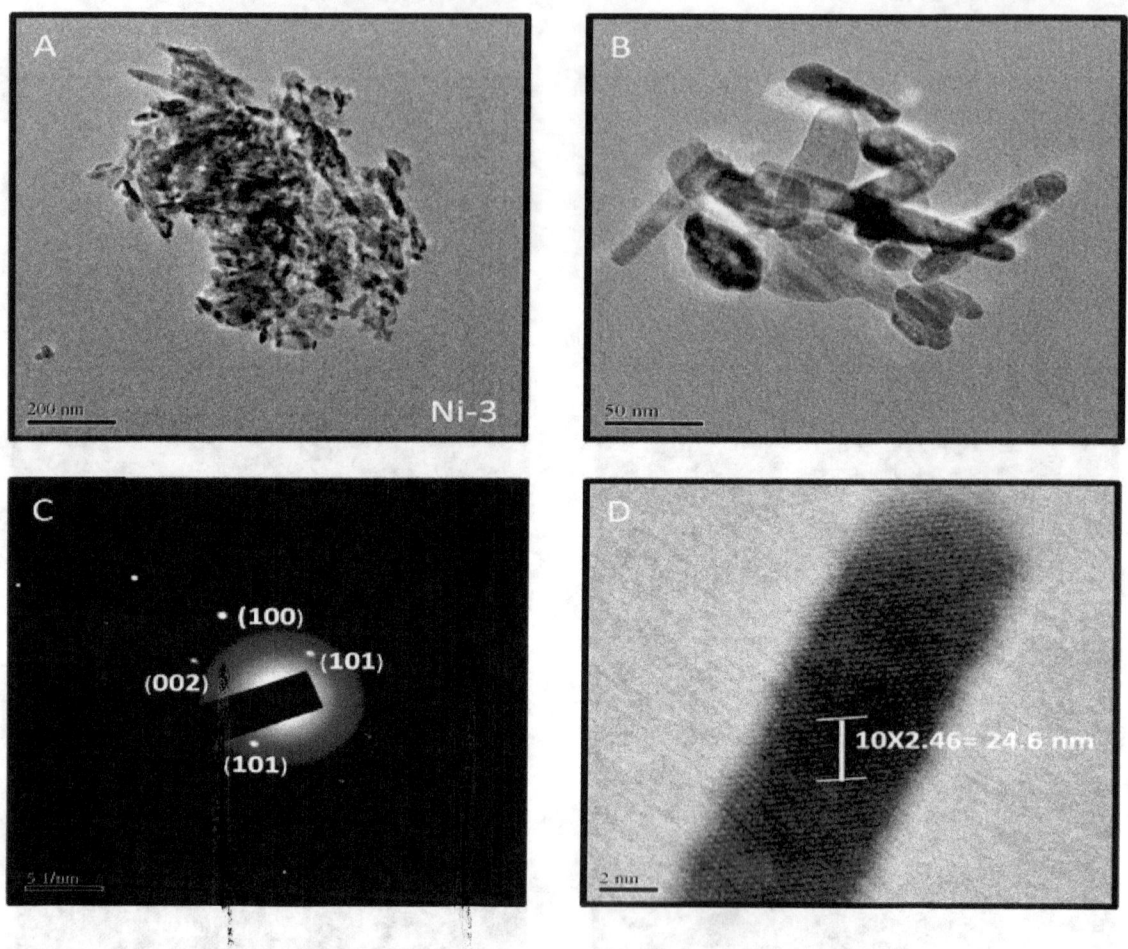

Figure 4.35: (A) TEM image of Ni-3 showing formation of elongated nanostructures, (B) High resolution TEM image of few Ni-3 nanorods, (C) SAED pattern focussed on an isolated nanorod as shown in fig. (D) with lattice spacing ~2.46 Å corresponding to the plane (101).

HRTEM images of Ni-3 sample are presented in figure: 4.35. Most of the nanostructures are agglomerated and elongated in nature (figure: 4.35A). Using standard software the length and diameters of few distinct nanorods were measured and average length and diameter were estimated 53 nm and 15.3 nm respectively. With increased resolution, clear formation of few Ni-3 nanorods were observed, which are shown in figure: 4.35(B). SAED pattern focussed on an isolated nanorod is presented in figure: 4.35(C), in which few lattice points corresponds to the diffracting planes (101), (002) and (100). The HRTEM image of an isolated nanorod with average length 25 nm and diameter 6 nm is shown in figure: 4.35D, the observed lattice spacing ~2.46 Å corresponds to the plane (101). It indicates the growth plane of the crystal is along (101) plane.

(c) Ni-5

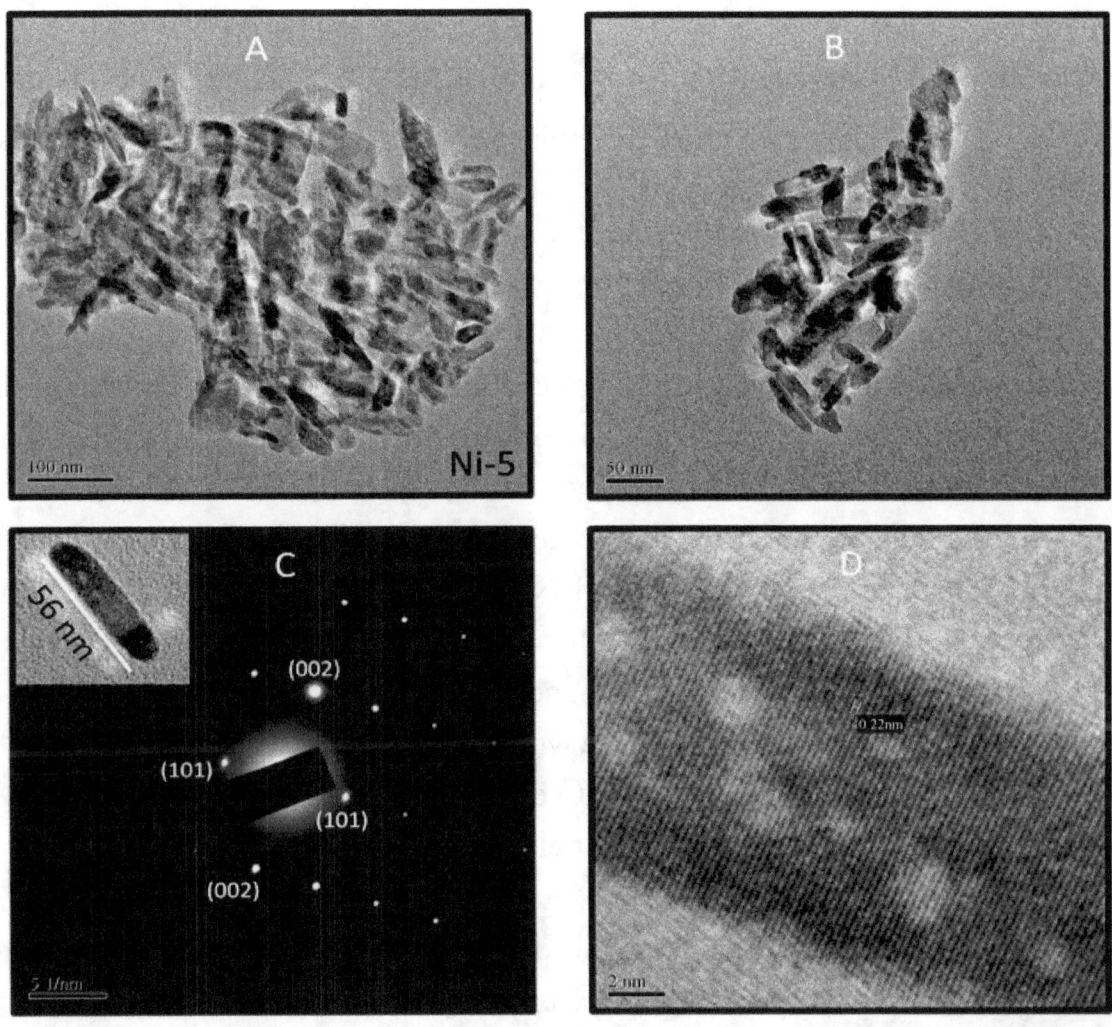

Figure 4.36: Figure (A) and (B) depicts formation of nanorods with av. length 44.5 nm and av. Diameter 12 nm. For the sample Ni-5, (C) SAED pattern, Inset: an isolated nanorod of av. length 56 nm and av. diameter 14 nm, (D) HRTEM image, when beam was focussed on the isolated nanorod; showing lattice spacing 2.2 Å,.

HRTEM study revealed the formation of nanorods with average length 44.5 nm and diameter 12 nm as observed in figure: 4.36A and 4.36B. Figure: 4.36C represents the SAED pattern focussed on the isolated nanorod of av. length 56 nm and av. diameter 14 nm (inset of figure: 4.36C). The lattice spacing 0.22 nm has been observed from the HRTEM image (figure: 4.34D) of the isolated nanorod as shown in the inset of figure: 4.36C. Clear lattice fringe pattern can be attributed due to the high crystalline nature of the as synthesised sample. In case of Ni-5, we have observed a large number of isolated nanorods. In this concentration of Ni the growth plane was confirmed along the (101) plane.

4.4 Significant observations

4.4.1 Structural

- It was observed that along with decrease in size, the strain also decreases. Thus, size and strain of the nanostructures can be simultaneously manipulated by changing doping concentration.

- HRTEM study for all samples exhibited clear fringe patterns; it was attributed due to high crystalline nature of the as-synthesised samples. The growth direction of the bare ZnO was found along (102), while for all TM doped samples it was found along (101) plane. In case of bare ZnO, we have noticed several lattice distortions on the surface of the nanocrystallites, this may play a significant role to influence optical and magnetic properties of the NSs.

- A significant change in lattice parameters (a and c) has been investigated from the structural study of the as-synthesised NSs. Increasing lattice parameters have been observed as a result of Mn doping. Out of three samples **Mn-3** doped ZnO showed highest values of lattice parameters (a and c).

- Pure ZnO and TM-doped ZnO have wurtzite structure and are further supported by FTIR. Undoped ZnO, TM doped ZnO exhibited similar FTIR broadening peaks due to stretching and vibration of different elements present.

- The identical IR peak for metal-oxide band stretching is observed near 500 cm^{-1} for all bare and TM doped ZnO. It was noticed that all samples exhibited IR peak near 500 cm^{-1}, but from the comparative FTIR plot it was evidenced that compared to other Mn:ZnO samples, **Mn-3** sample exhibited minimum %transmittance at ~500 cm^{-1} due to Zn-O and Mn-O stretching and bending. Observing the intense nature of the IR peak broadening for the sample Mn-3, it can be speculated that TM concentration can influence to shift IR peak positions.

4.4.2 Optical

- In the undoped ZnO specimen, mainly two asymmetrically broadened PL responses were observed: the first, in the UV region and the other in the green region. The UV-band is attributed to the near band edge emission, while the green band is reasonably attributed to the emission which has originated due to the intrinsic defect states of ZnO, such as oxygen vacancy, zinc interstitial and oxygen antisite. An associated lowering of intensity of the doped samples indicates suppression of radiative recombination process.

- In case of Co:ZnO system, the UV peak is red shifted and suppressed about 46% as compared with the bare ZnO sample, for the Co-3 doped sample the suppression decreases by about 48% and the peak width was broadened further. However, when Co-5 was doped no characteristic peak for near band edge emission was observed. This investigation indicates that Co concentration is playing significant role in displaying PL intensities in both UV and visible region.

- The suppressed emission peak as observed in the green region exhibited by the PL spectra of Ni:ZnO system may be attributed due to the presence of less defect in the Ni:ZnO crystal structure.

- In the UV- Visible study, it was observed that the sample Ni-3 exhibited highest transmittance (73%), while the Ni-1 sample exhibited the lowest transmittance (58%). The transmittance (%) for all samples in the visible region was recorded below 2.5%. It opens a wide possibility to use the sample in transparent TM:ZnO DMS.

- From the %Reflectance spectra of Co:ZnO system, it was noted that with increasing Co concentration the %Reflectance decreases.

- It has been observed that due to TM doping, the intensities of UV peaks in PL spectra get lowered and additional defect related peaks were generated. It reveals substantial influence in the PL intensities due to TM doping in ZnO.

ZnS Based DMS Nanostructures

Different aspects from my investigations on ZnS based DMS nanostructures will be highlighted in in this chapter. There have been reports on the fabrication and optical properties of semiconductor nanocrystals highlighting possible application in optoelectronic devices [284]. Among II-VI compounds, ZnS is a promising host material due to its thermal and environmental stability. Incorporation of both transition-metal ions and rare-earth ions in ZnS nanostructures by adopting chemical and physical techniques have been reported in recent years [285-287]. It has been used as base material for cathode ray tube luminescent materials [288-290]. In this research work, we highlight chemical synthesis of TM(Mn, Cr):ZnS nanostructures and their spectroscopic and structural characterizations. TEM and AFM micrographs confirm the formation of spherical nanostructures of TM doped ZnS. In case of Cr:ZnS, at lower magnification, unexpected growth of fractal-like patterns were confirmed though TEM micrographs, while at higher magnification these fractals were found to consist of individual nanoparticles of average size less than 30 nm. In the study, the possibility to control the fractal patterns by utilizing two different dielectric hosts (PVA and PVP) was depicted. Cubic crystalline structure of the as synthesised nano Cr:ZnS and Mn:ZnS samples was depicted from their XRD study. The UV-VIS spectrum of the colloidal ZnS:Cr nanostructures reveals that the lowest energy ZnS band gap excitation at 292 nm (4.25 eV) is higher than the bulk band gap (~3.67 eV).

5.1. Mn doped ZnS

We have adopted chemical synthesis route to fabricate Mn doped ZnS, which has been described in chapter-2. During synthesis Zinc chloride (aq.) and manganese chloride (aq.) solutions were prepared separately and then they were mixed together with varying Mn concentration from 0.008% to 5%. Out of nine different samples with varying Mn conc., the investigation results are presented for three samples with Mn conc. 0.25%, 0.188% and 0.008%, which were indexed as A, B and C.

(i) XRD characterization

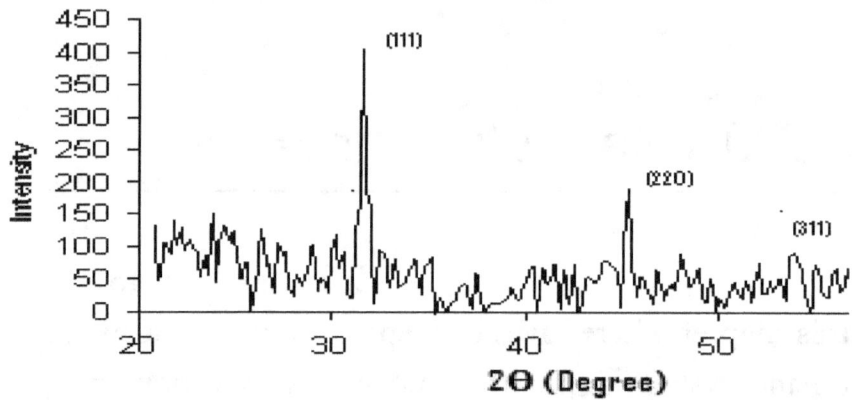

Figure 5.1: XRD pattern of Mn doped ZnS (sample-C), showing cubic crystalline structure of ZnS.

XRD study of all samples exhibits identical characteristic peaks. Figure: 5.1 shows the XRD pattern of the sample C. The study reveals cubic crystalline structure of ZnS corresponding to three diffraction peaks (111), (220) and (311). No additional phases due to Mn incorporation of ZnS were detected. Considering the intense diffraction peak (111), the average size of the nanostructures was estimated by using Scherrer's formula and found to be ~57nm.

(ii) UV- visible optical absorption study

Optical absorption study (OAS) plays an important role to observe size quantization effect in terms of blue shift and excitonic wavelength. Also, it ensures first hand signature of lowest possible transitions. In a study, it was claimed that for Mn:ZnS system, fabricated by thermal evaporation technique, the band edge absorption shifts from the higher wavelength region to lower wavelengths with increase in Mn concentration (0.04 %-0.12 %) [291]. In my case, (with conc. variation 0.008% to 0.25%), however, we observe the shifting in reverse order.

Figure: 5.2 represents UV-VIS spectra of nano ZnS:Mn samples. Long tailing in the wavelength range of OAS represents some inhomogeneity around the quantum size distribution. Earlier works on ZnS:Mn have showed the excitonic absorption band at ~323nm and absorption related to Mn- impurity state at ~334nm [292]. In my study, we have noticed sharp excitonic absorptions between 330 nm-280 nm, corresponding to samples A, B and C, (with Mn concentrations 0.25 %, 0.188 %. And 0.008 %), respectively. The absorption was also found to decrease along with Mnconcentrations. The corresponding blue shifted energy of the nanoparticles lies between 3.75 eV to 4.43 eV (Table: 5.1). This result shows absorption peak is blue shifted from the bulk band gap ~3.68 eV of ZnS. Significant energy gap enhancement as observed in the optical absorption spectra reveals quantum confinement of the particles due to reduction of their sizes.

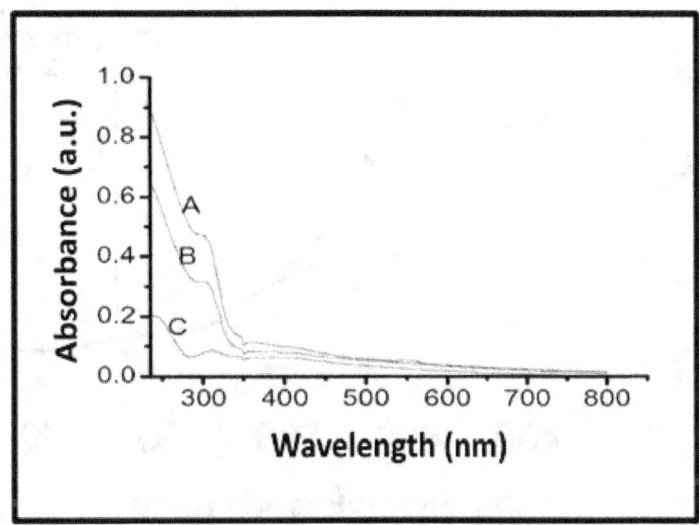

Figure 5.2: UV-Visible optical absorption study of the Mn doped ZnS (sample-A, B and C).

Table 5.1: Absorption peaks and enhancement of band gap energies corresponding to the samples A, B and C.

Sample	Mn Conc. %	Abs. Peak	E_g (eV)	Enhancement energy (eV)
A	0.250	330	3.75	0.07
B	0.188	315	3.94	0.26
C	0.008	280	4.43	0.75

(iii) Photoluminescence study

Photoluminescence (PL) study plays an important role for investigating discrete energy levels of quantum confined solids. PL study of the ZnS:Mn sample was carried out at ~300K with an excitation wavelength of 325nm is presented in figure: 5.3. Analysis of the PL study of the samples revealed that emission centred at around 420 nm can be attributed due to the trap emission related to ZnS nanostructures. A weak emission band ~ 560-580 nm can be ascribed to the d-electron transfer from Mn^{2+} to the host ZnS lattice. Previously, it was established that

Mn doping onto ZnS host could give out orange-yellow emission [144]. All samples show nearly identical characteristic peaks. The broadened but significantly suppressed peak due to Mn incorporation in the ZnS lattice is sown in the inset of Figure: 5.3.

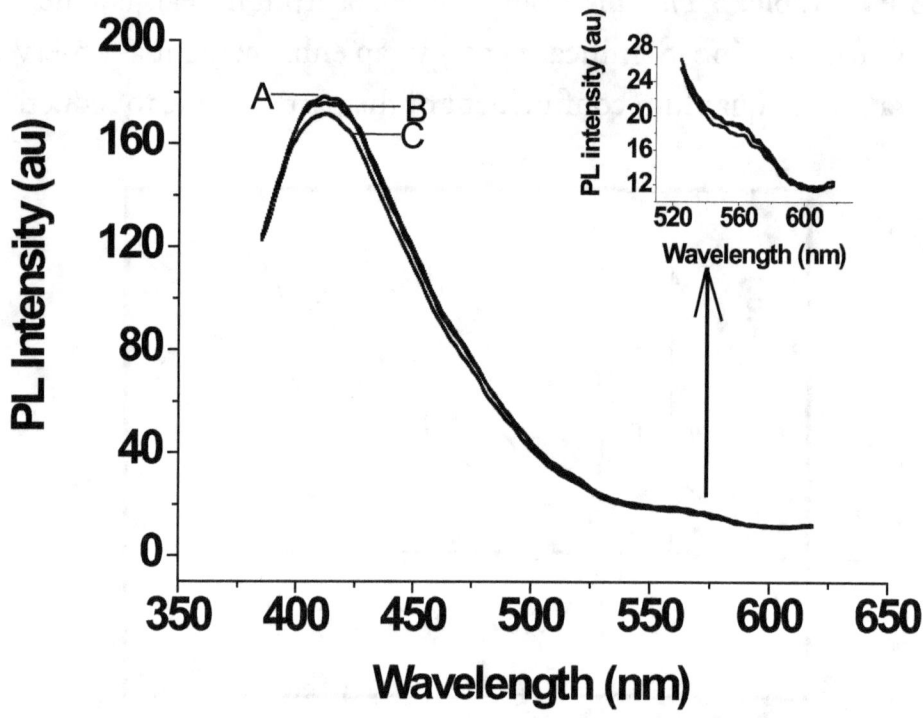

Figure 5.3: Photoluminescence study of the Mn doped ZnS (samples-A, B and C).

In general, the photoluminescence spectra of ZnS nanostructures show a strong blue luminescence, peaking at ~420 nm and having an extended tail. The highly asymmetric and broadened emission spectra around 400nm-450nm indicate the involvement of number of luminescence centres that corresponds to different kinds of defects. These defects can be interstitial, lattice vacancy and impurity related types. The radiative transitions via all these states can be superimposed which result in a broad peak centred at ~420 nm.

Basically, both the Schottky and Frankel defects exist in all crystalline solids, but one type of defect is dominant at any time, since they are associated with different energy of formation.

It was known that in cubic ZnS crystal the Schottky defects are more dominant. Primarily, Schottky defects of ZnS involve following vacancy related defects-

\# S^{2+} vacancies ($V^{\bullet}_s$),

\# Zn^{2+} vacancies ($V^{\bullet}_{zn}$),

In case of ZnS nanostructures, surface S^{2+} vacancies ($V^{\bullet}_s$) are more dominant over other defects. In ZnS nanoparticles, 'S vacancy ($V^{\bullet}_s$)' and 'Zinc interstitial (Zn_i)' behave as 'shallow donors' (electron traps), on the other hand, 'Zn vacancy ($V_{zn}^{\cdot}$)' and 'S interstitial (S_i)' act as 'deep acceptor' (hole traps) levels. For this, sulphur vacancies ($V_s^{\cdot}$) generate localized donor sites just below the conduction band. Excitation of these produce a positive charge and conduction band electron. The localized charge exerts a potential, which can further trap electrons. Thus, upon excitation, Sulphur vacancies ($V_s^{\cdot}$) are pumping the electrons into conduction band. Emission occurs when a captured electron recombines with a hole [293].

(iv) TEM study

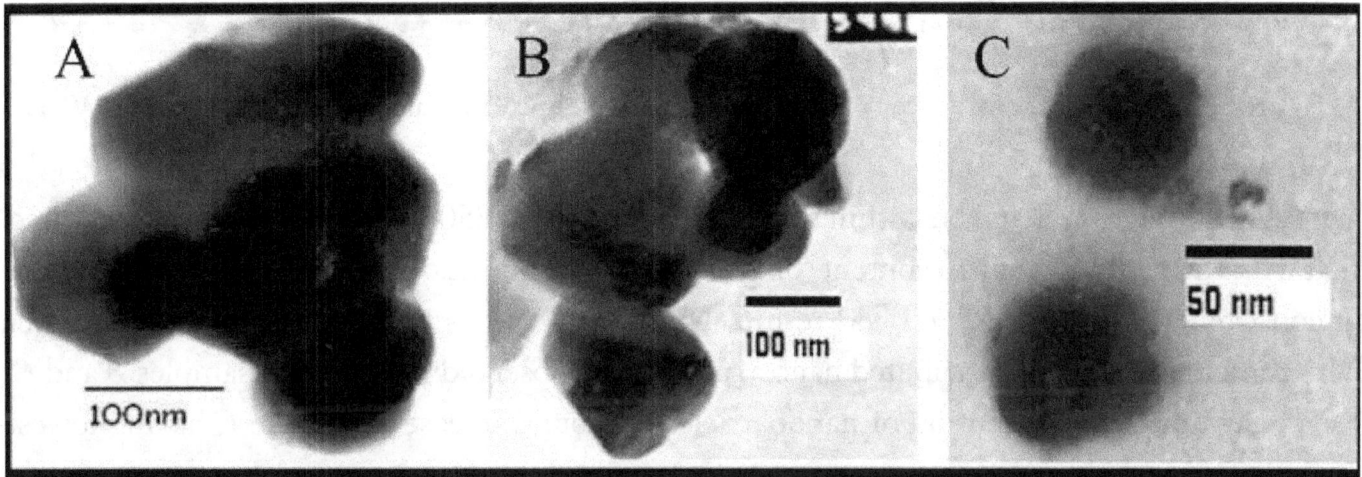

Figure 5.4: TEM measurement of Mn doped ZnS nanostructures (a) A(0.25%), (b) B(0.188%), (c) C(0.008%).

Transmission electron microscopy study was performed for all the samples. The study has revealed that with decrease in Mn-concentration, the size of nanoparticles decreases and clustering effect tends to cease. The average sizes for samples A, B, and C are found to be ~ 80 nm, 75 nm and 60 nm. A mixture of cubical and spherical nanostructures was seen in the micrograph of sample A. For sample C, with Mn concentration 0.008% the average size calculated is in good agreement with the XRD result. The sample B exhibited nearly spherical nanostructures, while C gave signature of isolated nanospheres. Looking at arrangement of nanoparticles, one can infer that changing Mn contents can also influence nature of clustering along with shape effect.

(v) AFM studies

The surface morphology of all the three samples casted on glass substrates were studied by using atomic force microscopy. A typical AFM micrograph for the samples *A*, *B* and *C* is presented in Figure: 5.5.

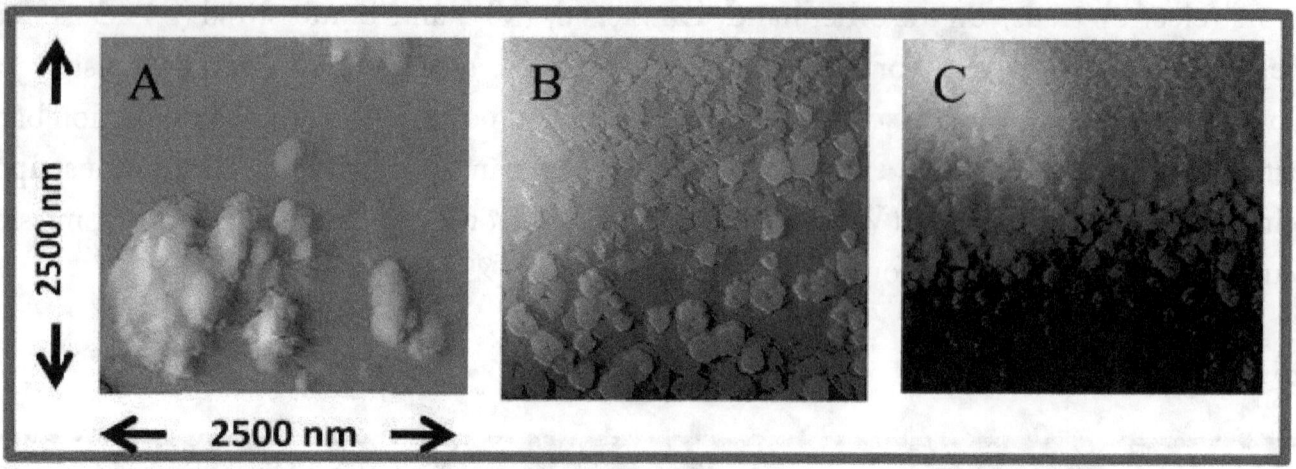

Figure 5.5: Atomic force microscopy (AFM) image of Mn doped ZnS nanostructures (A) 0.25%, (B) 0.188%, (C) 0.008%.

The AFM images were taken within specimen area 2500 X 2500 sq. nm. It has been observed that with the reduction of Mn content (0.25% to 0.008%), the size of the nanoparticles decrease in consistency with my earlier TEM results. More clustering is seen for sample *A* with larger Mn-content. Uniformly distributed nanostructures are observed for both the samples *B* and *C*. Relatively uniform distribution of nanoparticles are found in case of sample *C* having lowest Mn-content.

The present work has revealed that optical, spectroscopic and structural changes occurring in ZnS:Mn nanoparticle system can be studied efficiently. Significant blue shifts in the absorption edge, as observed in the OAS, are ascribed due to the reduction of nanoparticle size. PL study shows an orange-yellow band, confirming incorporation of Mn in the ZnS host lattice. The average size of the nanoparticles estimated from TEM and XRD studies are almost identical. Supporting results from UV-Vis study, TEM and AFM indicate that there is a tendency of decreasing particle size with decrease in Mn concentration.

5.2 Cr doped ZnS

There is a growing interest and continuous demand in the fabrication of self similar materials, parts, features etc. for application in nanotechnology. One of the difficult tasks was to produce materials that incorporate multiple length scales simultaneously, from nano and micro, to micro scale. In fact, fractals could achieve this goal because of their self similarity and sustainability in the real world as well as in theory [294,295]. Fabrication of micro structured fractals requires self assembly since direct manipulation of such self similar structures at the lower end of the scale is extremely difficult to achieve [296]. In this work, we report physico-chemical synthesis of Cr:ZnS nanostructures and their spectroscopic and structural characterizations.

(vi) XRD study

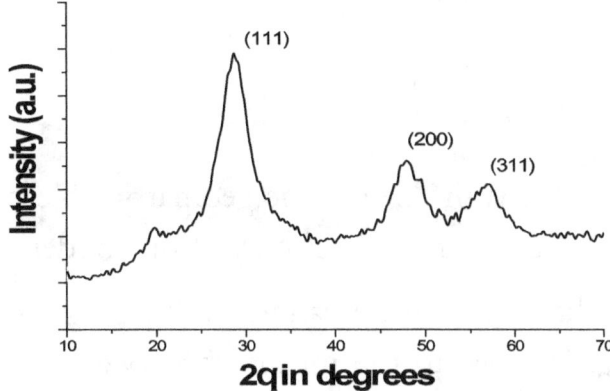

Figure 5.6: XRD pattern of Cr doped ZnS exhibiting cubic crystalline structure of ZnS corresponding to three diffracting peaks (111), (200) and (311).

The x-ray diffraction pattern of nano ZnS:Cr in PVP is shown in Fig.1(e), depicts cubic crystalline structure corresponding to three diffraction peaks (111), (220) and (311). The broad XRD diffraction peaks suggest the existence of nanostructures. The XRD pattern gives an idea of rough estimation of average particle size ~15 nm, obtained by measuring full-width-at half maxima (FWHM) using Scherrer formula.

(vii) TEM study

At lower magnification of TEM study unexpected growth of fractal-like patterns [294,295] were confirmed though TEM micrographs (Figure: 5.7A), while at higher magnification these fractals were found to consist of individual nanoparticles of average size less than 30 nm (Figure: 5.7B). In figure: 5.7C, distribution (%) of nanoparticles with particle size in a 0.585 sq. µm area is shown. It is observed that the maximum number of particles lie within the range of 10 nm to 20 nm.

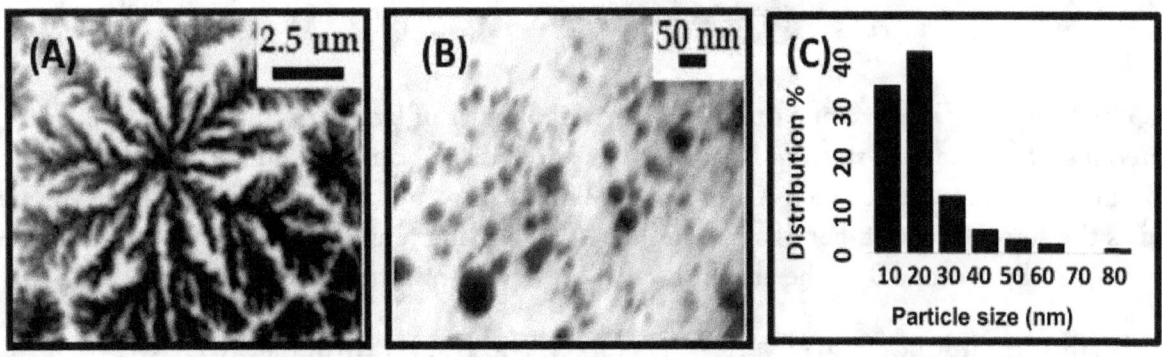

Figure 5.7: TEM micrographs (A) Fractal pattern at higher magnification, (B) spherical nanoparticles of Cr:ZXRD pattern of Cr doped ZnS exhibiting cubic crystalline structure of Zn:SnS in PVA matrix at closer inspection, (C) distributioon of particles with average grain suize.

The fractal like feature was further confirmed by systematically measuring its fractal dimension, typically defined by the divider formula [297]

$$D_f = \lim_{r \to 0} \frac{\log N(r)}{\log(r)} \qquad \to 5.1$$

Where r is the length unit, $N(r)$ is the size of the geomatric object measured with unit r. To measure D_f, We used a methode illustrated by Andrea Lomannder et. al. [296].

For a given value of r, the ruler usually cannot walk along the contour exactly with an integer number of steps, giving rise to the mismatch between the start and the endpoints (points A and B in figure: 5.8). In such cases, N was calculated as the sum of the number of full steps and the fractional length between the start and the end points with respect to r. In the example shown in figure: 5.8(a), r =125 µm, it looks 10 steps (thick line) starting from A to proceed to the point B along the contour. The remaining distance between A and B (thin line) is 50 mµ, whose fractional length with respect to r is 50/125 = 0.4, thus the total contour length is 10.4. In figure: 5.8(b), r = 50µm, the contour length yeilded an average value of N=38.75. The measured value of N depended on the location of the starting point of the contour. Inspired by the Nyquist sampling theorem [298], several values of N starting at different points on the contour were measured and averaged.

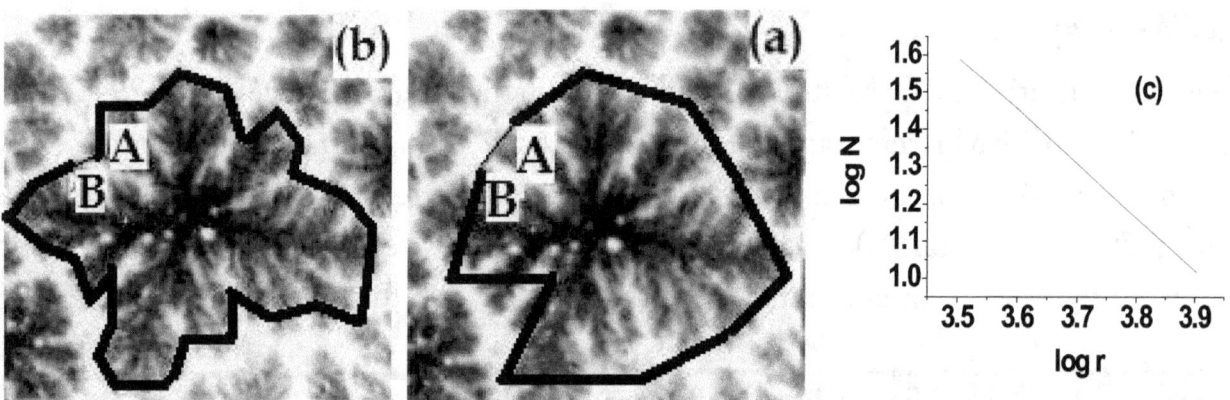

Figure 5.8: Analysis of the fractal patterns, (a) A straight rular of length r (thick line), walks along the nanomaterial surface contour, starting from point A, and ends at point B, Left: r= 125 µm, and in figure (b) r = 50 µm, The distance between A and B (thin line) is measured to be d. The path length N is then defined as the sum of the number of steps and d/r. (c) A log-log graph of N versus r (µm). The fractal dimension, D_f is the absolute value of the slope, in this case, D_f= 1.48.

The graph (figure: 5.8c) of log[r] vs. Log[N] gave the fractal dimension D_f= 1.48. This value corresponds to the fractal dimension D_f of a viscous fingering system [299], whose formation mechanism is known as diffusion-limited aggregation (DLA) [300,301]. In the plane, D_f = 1.71; however, in real systems, depletion effects may decrease D_f from this ideal value to 1.4 [301,302]. Thus, it can be interpreted that the fractal like feature was formed through DLA-like process. Change of pH and the presence of disulphide bond are important for the formation of fractals through DLA-like process [296]. Molecular self- assembly takes place via a subtle balance between non covalent bond interactions that result in the formation of well-defined structures [302-307]. The cysteine disulphide bond should remain stable at lower pH [308,309]. In a similar work [139], The authours have shown production of Cr:ZnS nanoparticles without formation of fractals. We have followed the same method; but, only adopting the replaced treatment of H_2S by acidic Na_2S treatment. So, we expect that, in my case the presence of excess disulfide bond could be responsible for the formation of fractals.

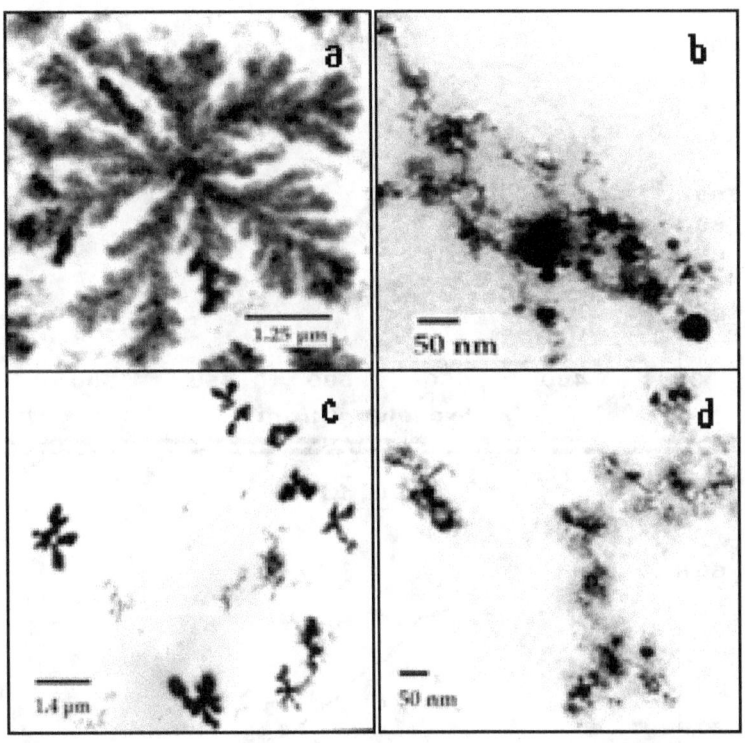

Figure 5.9: TEM micrograph (a) fractal pattern observed for Cr:ZnS system embedded in PVA, (b) high resolution TEM image confirms formation of spherical nanostructures, (c) TEM image of Cr:ZnS system embedded in PVP, (d) high resolution TEM image when Cr:ZnS system is embedded with PVP.

Further study of the formation of fractal structures with Cr:ZnS system was extended by considering another dielectric host PVP. For this, we have prepared Cr:ZnS sample by adopting same chemical route only by considering both dielectric host, PVA and PVP separately. My interest was to find out any change in TEM micrographs of fractal pattern due to the change of dielectric hosts. TEM characterizations were carried out carefully for higher and lower magnification. In both cases we have observed formation of fractal structures at lower magnification while at higher magnification existence of spherical nanoparticles was confirmed. ZnS:Cr nanostructures in both dielectric hosts PVA and PVP organize in some specific way in the form of fractal (figure: 5.9a and 5.9c). At higher magnification the existence of ZnS:Cr spherical nanostructures were observed in these fractals (figure: 5.9b and 5.9d).

(viii) Photoluminescence study

The photoluminescence spectra of bareZnS and ZnS:Cr nanostructures (fractals) was performed at room temperature with excitation wavelength 325 nm and presented in figure: 5.10. The PL spectra of bare ZnS shows a prominent peak at 424 nm, which is attributed as band edge emission peak of ZnS.

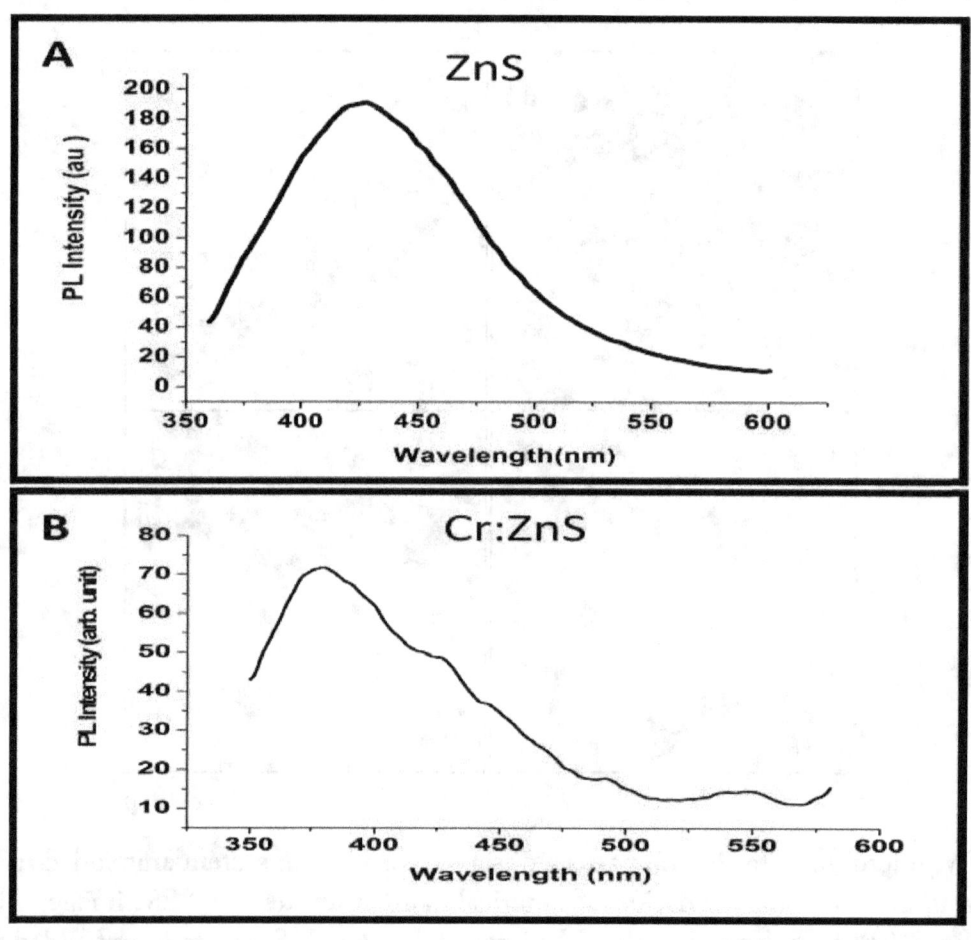

Figure 5.10: Photoluminesce spectra of (A) bare ZnS system, (B) Cr doped ZnS embedded in PVA.

The Cr doped ZnS system exhibited blue shift of this band edge peak and found at ~380 nm, which can speculated due to the reduction of particle size in the Cr:ZnS system. The additional peaks around 425 nm and 490 nm are due to the trap states arise as a result of sulphur vacancy and Zinc vacancy respectively in the nanostructures. Some less prominent peak is also observed near 550 nm, which also arises due to some trap states as a result of doping of Cr in the ZnS nanostructures. Identical peaks are observed for PVP matrix along with higher luminescence effect.

(ix) UV-Vis spectroscopy study

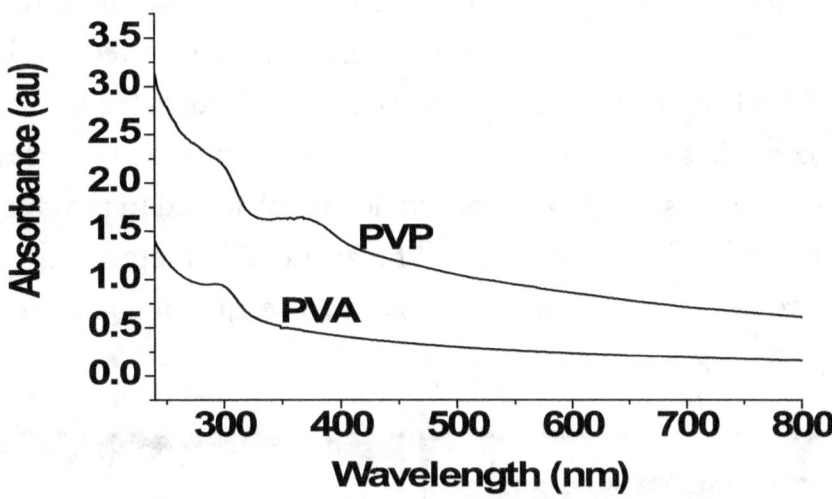

Figure 5.11: UV-Vis. spectra of Cr:ZnS system embedded in PVA and PVP

The overview electronic absorption spectrum of the colloidal ZnS:Cr^{2+} nanostructures in PVA and PVP dielectric hosts are presented in figure: 5.11. The lowest energy ZnS band gap excitation at 292 nm (4.25 eV) in both dielectric hosts is higher than the bulk band gap (~3.67 eV). This shows clear blue shift due to quantum confinement effect.

(x) EDX study

Energy Dispersive X-ray spectroscopy (EDS) study of the Cr:ZnS system embedded in PVA as presented in figure: 5.12 shows the existence of Cr, Zn, and S in the samples. The study reveals that the content of S, Zn and Cr in the ZnS:Cr sample embedded in PVA are 42.19 %, 55.36% and 2.45% respectively.

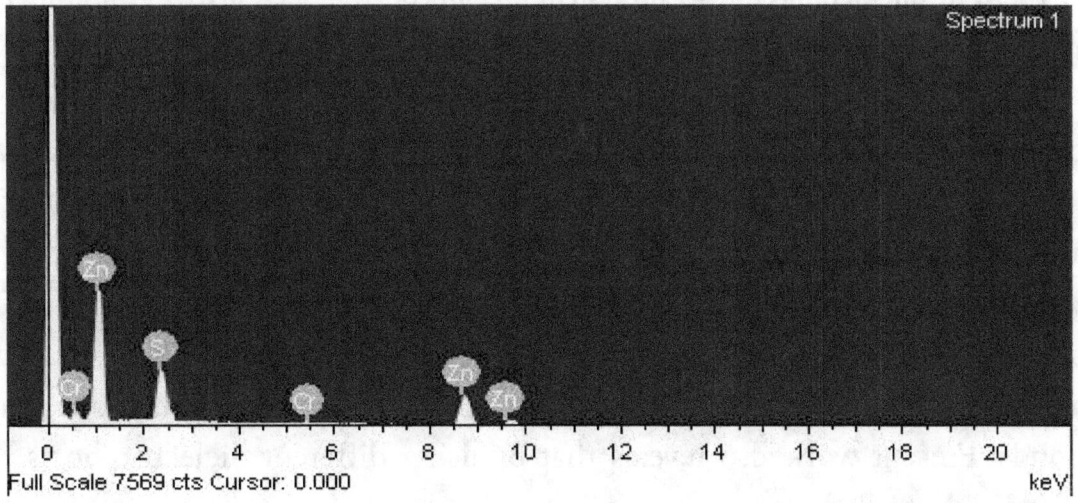

Figure 5.12: EDX spectra of Cr:ZnS system embedded in PVA.

(xi) AFM study

The surface morphology of the samples casted on glass substrates were studied by using atomic force microscopy (AFM). The typical AFM micrographs on selected section of the ZnS:Cr sample in PVA and PVP dielectric matrix are shown in figure: 5.13(a) and 5.13(b). AFM image in PVA matrix depicts uniform surface morphology with surface area 2.5μm x 2.5 mμ, while uniform distribution of spherical nanoparticles are observed in the coloured AFM image of Cr:ZnS embedded in PVP dielectric host. The average diameter of the spherical Cr:ZnS nanostructure was estimated as ~14 nm, which is in close agreement with the size estimated from XRD peak broadening.

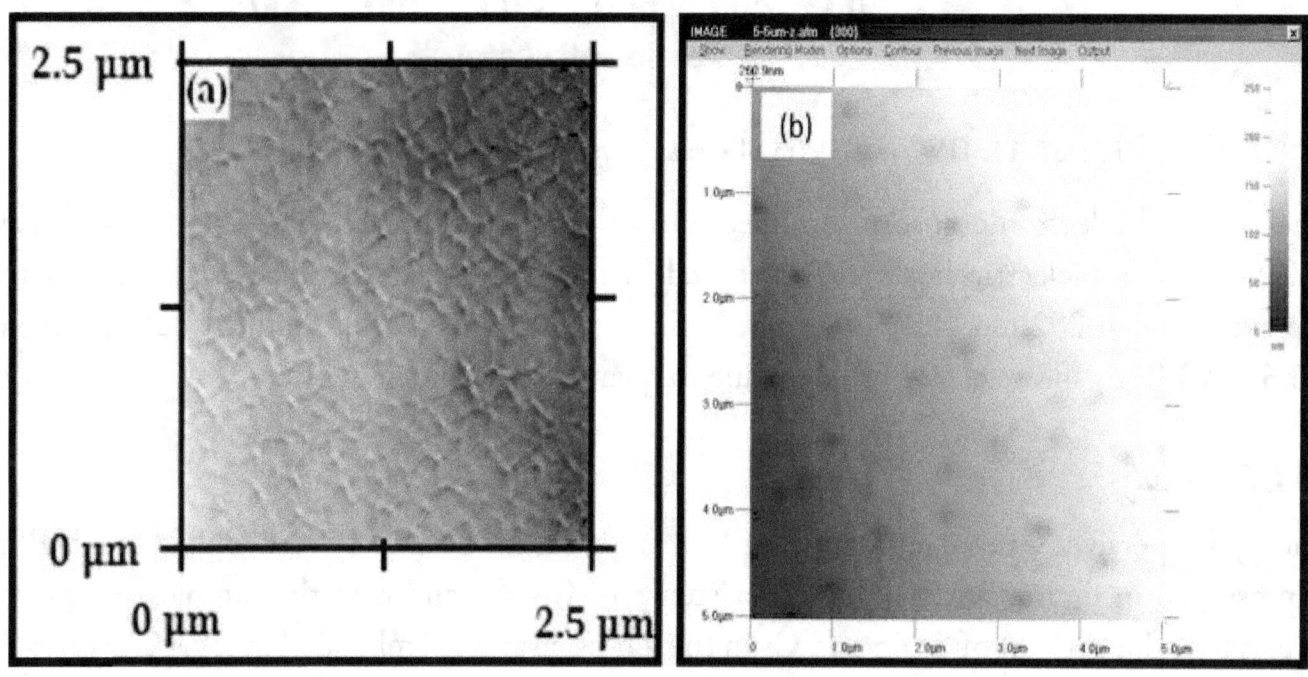

Figure 5.13: AFM micrographs of Cr:ZnS system (a) embedded in PVA matrix exhibiting uniform surface mophology with area 2.5μm x 2.5μm, (b) embedded in PVP matrix showing spherical nanoparticles distributed within area 5μm x 5μm (coloured image).

In this work we have reported generation of fractal patterns while fabricating Cr doped ZnS nanostructures embedded in dielectric hosts by adopting inexpensive chemical technique. Photoluminescence shows emission peak at ~ 550 nm due to incorporation of Cr^{+2} ions into ZnS host. From the calculation of fractal dimension as 1.48, it is evident that the development of fractals in the samples is due to diffusion limited aggregation (DLA). The well defined fractal structure is a molecular self assembly arises due to balanced interactions among non-covalent bonds. Present work also reveals that by using different dielectric hosts the fractal structure may be controlled.

5.3 Significant observations

5.3.1 Structural

- The XRD study of Mn and Cr doped ZnS reveals cubic crystalline structure of ZnS corresponding to three diffraction peaks (111), (220) and (311). No additional phases due to Mn and Cr incorporation of ZnS were detected. From XRD data the average size of the nanostructures was estimated to be ~57 nm for Mn:ZnS system, while ~15 nm for Cr:ZnS system.

- From the TEM study of the Mn:ZnS system it has been concluded that by changing Mn content on can influence the shape, size and clustering nature of the NSs. For higher Mn content, a mixture of cubical and spherical; for medium Mn content, nearly spherical and for lower Mn content isolated NSs were seen in the TEM micrographs. The sizes of the NSs as calculated from XRD data is in good agreement with the TEM results.

- TEM study of Cr:ZnS system revealed unexpected growth of fractal-like patterns at lower magnification, while at higher magnification these fractals were found to consist of individual spherical nanoparticles. Further study of showed that these fractal patterns can be menipulated by changing dielectric hosts. It was confirmed that the development of fractal patterns is due to diffusion limited aggregation.

- AFM studies performed on TM (Mn, Cr) doped ZnS systems exhibited uniform surface morphology of the NSs. Sizes of the NSs estimated from AFM micrographs is in good agreement as calculated from XRD and TEM analysis.

5.3.2 Optical

- In the UV-Visible study of the Mn:ZnS nanoparticles, it was noticed that the corresponding blue shifted energy lies within the range 3.75 eV - 4.43 eV. This result shows absorption peak is blue shifted from the bulk band gap ~3.68 eV of ZnS. Significant energy gap enhancement as observed in the optical absorption spectra both Mn:ZnS and Cr:ZnS systems reveals strong quantum confinement of the particles due to reduction of their sizes.

- Room temperature PL study of the bare ZnS and Mn, Cr doped ZnS systems was carried out with an excitation wavelength of 325nm. The revealed that emission centred at around 420, nm, which can be attributed due to the trap emission related to ZnS nanostructures.

- The highly asymmetric and broadened emission spectra Mn:ZnS system around 400nm-450nm indicate the involvement of number of luminescence centres that corresponds to different kinds of defects. These defects can be interstitial, lattice vacancy and impurity related types. The radiative transitions via all these states can be superimposed which result in a broad peak centred at ~420 nm.

- For Mn:ZnS system, a weak emission band ~ 560-580 nm, while for Cr:ZnS, a broadened but significant peak around 550 nm were observed. It can be ascribed to the d-electron transfer from Mn^{2+} and Cr^{2+} to the host ZnS lattice.

Magnetic Measurements

Magnetic measurements of the as fabricated samples were carried out by utilizing magnetic force microscopy (MFM) and super conducting interference device (SQUID). For MFM measurements, Mn and Cr doped ZnS samples were coated on glass substrate with area 1cm x 1cm. For MFM characterizations samples were sent to Centre for Microscopy Microanalysis & Image Processing, University "Politehnica" of Bucharest, Romania. For SQUID measurements, pallets were prepared from the powdered transition metal (TM) doped ZnS and ZnO samples (TM: Co, Mn, Cr, Ni). All SQUID measurements were carried out at UGC, DAE consortium Indore. The samples were characterized within 1-2 weeks from their synthesis.

6.1 Magnetic measurement of bare ZnO

The magnetic characterization of undoped ZnO was carried out by using SQUID. To explore magnetic properties of bare ZnO and TM(Mn, Co, Ni) doped ZnO nanostructures we have performed SQUID measurements in two ways:

a. M~T response: Magnetic moment (M) of the samples was recorded by varying temperature (T) within the range 4K to 300K. During M~T study, investigation was done for both zero field cooling (ZFC) and field cooling (FC). ZFC and FC was measured for two different fields, 100 Oe and 0.1 T.

b. M~H response: By varying magnetic field magnetic moment exhibited by the samples were recorded. The magnetic field was varied from zero to several Tesla in both directions.

The M~H curves of bare ZnO exhibited typical diamagnetic nature. The M~H response of bare ZnO is presented in figure: 6.1. The diamagnetic behavior of ZnO arises due to the paired electrons of its d orbital, which is responsible for the absence of a permanent net magnetic moment per atom. Then, when electrons are paired together, their opposite spins cause the magnetic fields to cancel with each other. Accordingly, when an applied magnetic field is acting on this atom slightly unbalances their orbiting electron and creates small magnetic dipoles within the atoms which oppose the applied field. This action is responsible to produce a negative magnetic effect.

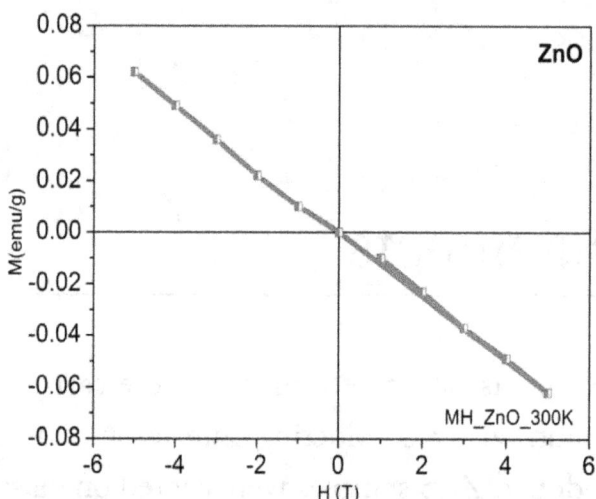

Figure 6.1: Dependence of magnetization with applied magnetic field, M~H response of bare ZnO at 300K.

6.2 Magnetic measurement of TM doped ZnO

(i) Mn doped ZnO DMS

Magnetic characterization of the Mn:ZnO samples were carried out by utilizing SQUID within a temperature range of 4- 300 K. Initially, we have investigated the change of magnetic moment (M) of the sample with respect to the temperature (T). The magnetic moment of the samples was recorded by varying temperature within the range 4K -300K for both at zero field cooling (ZFC) and field cooling (FC). ZFC and FC curves give the first hand signature of magnetic response of the specimen. The results of Mn:ZnO samples are being presented in figure: 6.2. M~T response for Mn-1 and Mn-5 samples are presented in Figure: 6.2A and 6.2C. For these two samples the sharp magnetic moment below 50 K was recorded beyond 0.0014 emu while for the sample Mn-3 the magnetic moment was recorded beyond 0.006 emu (figure: 6.2B). The M~T response of Mn-3 sample in a narrow range of temperature is presented in figure: 6.2D. The significant difference between FC and ZFC curve as indicated in the figure: 6.2D, which indicates the existence of ferromagnetism in the sample where Curie temperature is above 300 K. The overall M~H for Mn-1 sample in a wide range of magnetic field at 10K is shown in figure: 6.3A. The saturation magnetization (M_s) exhibited by Mn-1 at

10K was recorded as 0.1029 emu. The M~H response at 10K for the sample within -4000 Oe to +4000 Oe is shown in figure: 6.3B. The remanence magnetization (M_r) and coercive field (H_c) for Mn-3 at 10K were recorded as 0.00418 emu and 1200 Oe respectively.

For practical applications, room temperature ferromagnetism is a vital point to be considered. This is one of the major objectives of my investigations. We have studied variations of magnetic moments (M) of Mn:ZnO system with respect to the change of magnetic field (H) at room temperature. The M~H response for all samples (Mn-1, Mn-3 and Mn-5) at 300K is presented in figure: 6.4. All samples exhibited hysteresis loop indicating ferromagnetic nature of the Mn:ZnO system at room temperature.

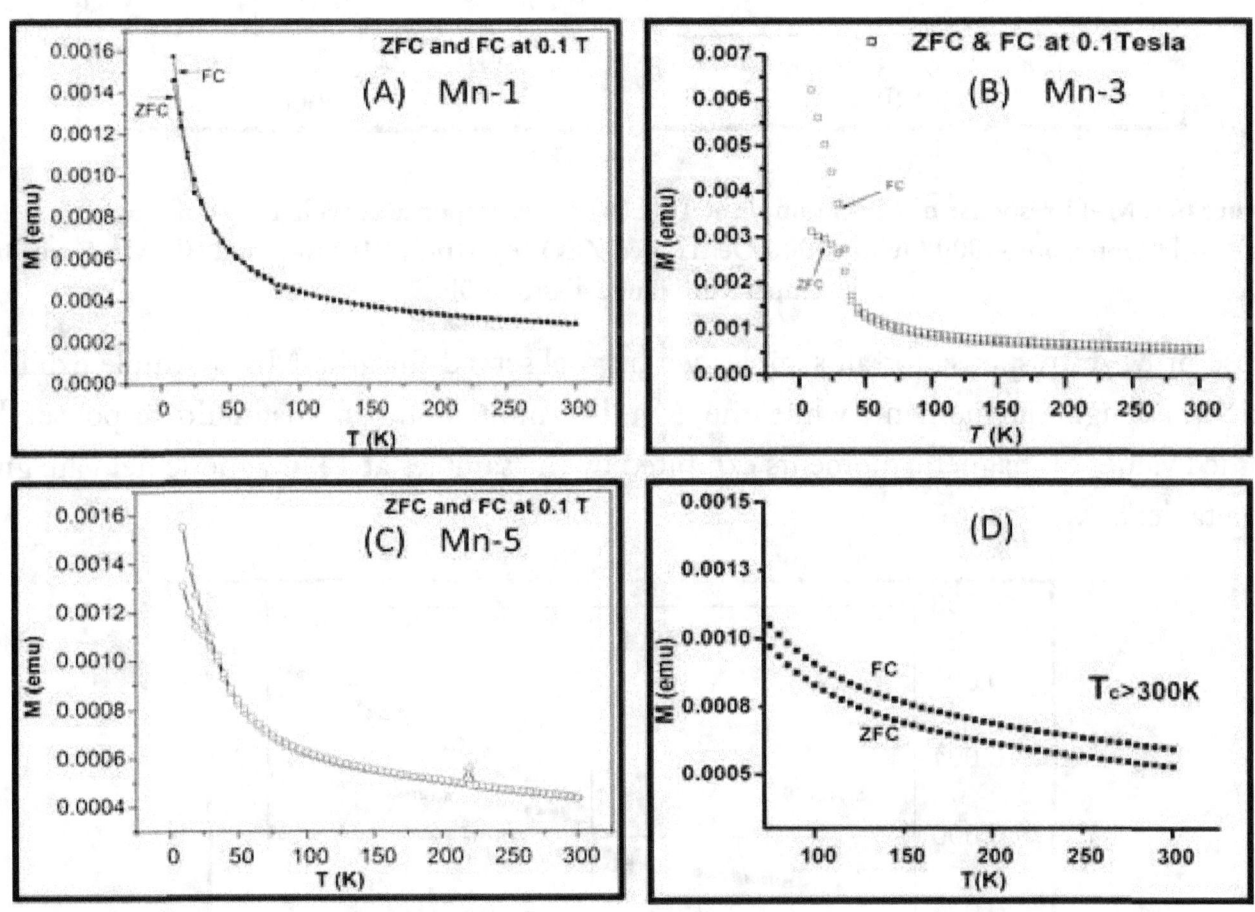

Figure 6.2: M~T response (ZFC and FC at 0.1T) of Mn doped ZnO (A) Mn-1, (B) Mn-3 and (C) Mn-5 within temperature range 4K to 300k, (D) ZFC and FC for Mn-3 sample within narrow temperature range (75K to 300K) showing significant difference between ZFC and FC curves.

It can be noted that Mn concentration plays an important role to influence magnetic moment in the specimen. As we have observed from ZFC and FC responses for the samples (figure: 6.2), Mn-3 sample exhibited highest magnetic moment compared to other two samples. Since, Mn-3 sample contains intermediate Mn concentration, so control over concentration of transition metal at the time of doping is a vital point in having considerable magnetic response.

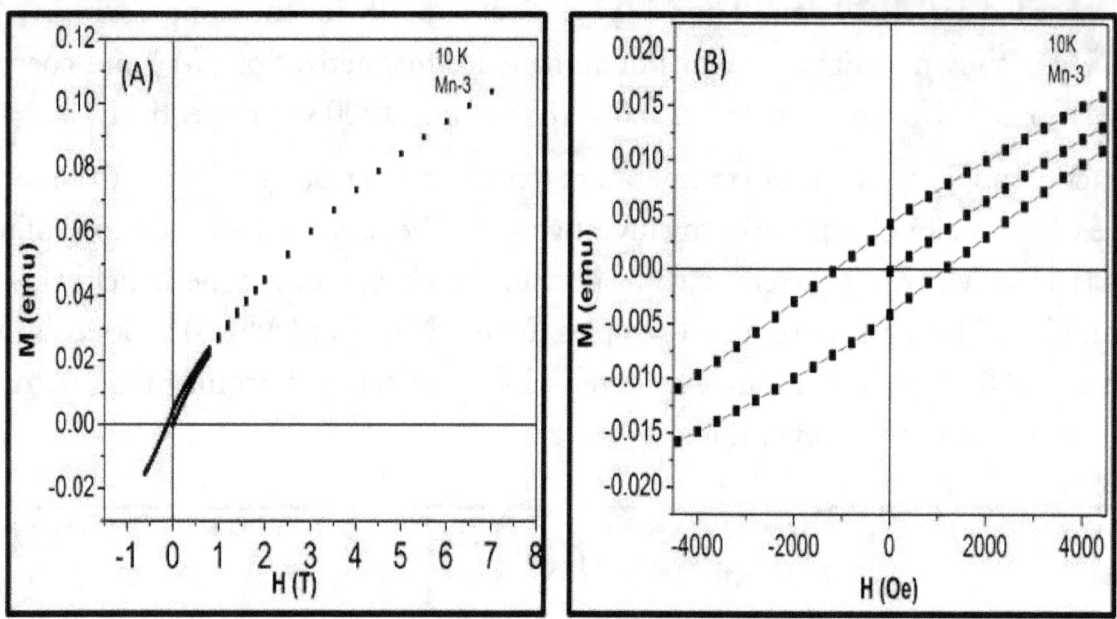

Figure 6.3: M~H response of Mn-3 sample at 10K (A) M~H response at wide range of magnetic field, (B) M~H loop from -4000 Oe to +4000 Oe. Doped ZnO (A) Mn-1, (B) Mn-3 and (C) Mn-5 within temperature range 4K to 300k.

In case of M~H response for all samples we have observed that the Mn-3 sample exhibited maximum magnetic moment, while the Mn-1 exhibited lowest magnetic response. The recorded values of magnetic moments exhibited by the samples at 1T magnetic field are given in the table: 6.1.

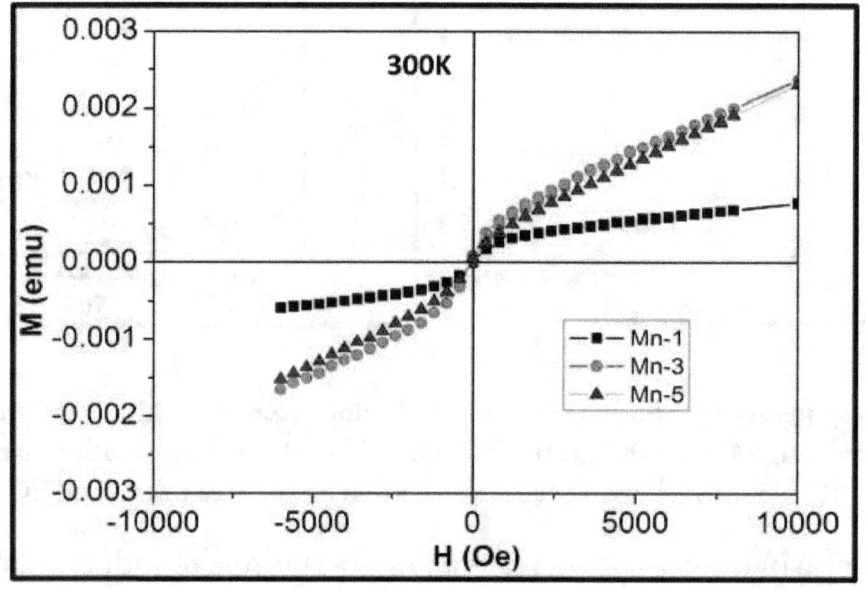

Figure 6.4: M~H response of Mn-1, Mn-3 and Mn-5 at 300K, showing hysteresis loop for all samples.

Table 6.1: **Magnetic moment exhibited by Mn doped ZnO samples at magnetic field 1T.**

Sample	Magnetic moment (emu)
Mn-1	7.62×10^{-4}
Mn-3	2.35×10^{-3}
Mn-5	2.30×10^{-3}

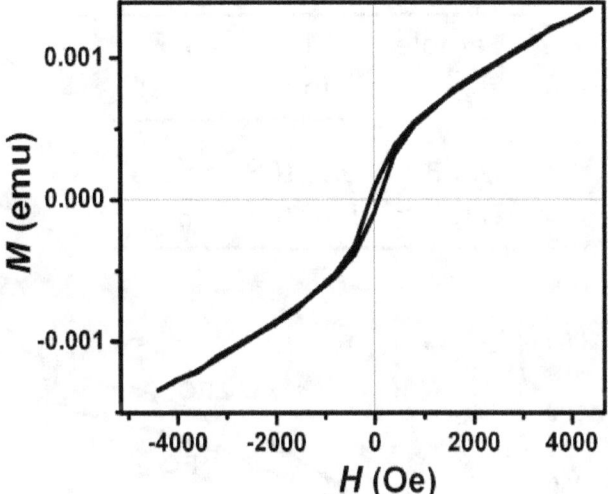

Figure 6.5: M~H response of Mn:ZnO (Mn-3) exhibiting hysteresis loop at 300K.

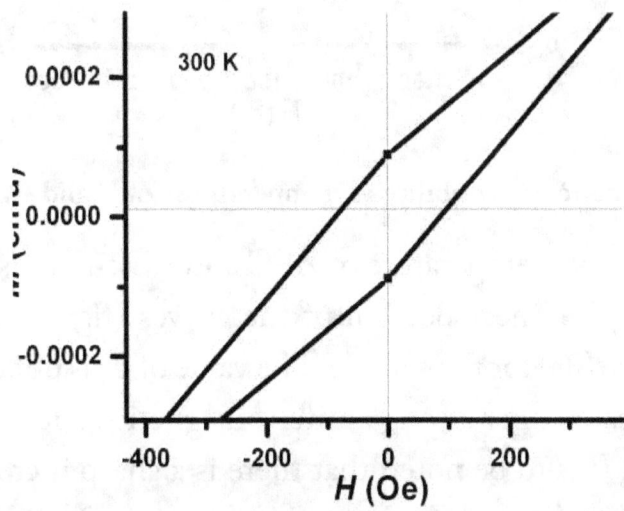

Figure 6.6: M~H response of Mn:ZnO (Mn-3) at 300K in a selective mag. field showing enlarged hysteresis response. at the origin.

For better understanding of the magnetic response exhibited by the Mn-3, the M~H loop at 300K is presented in figure: 6.5. The selective response of M~H curve at the origin for the sample

Mn-3 at 300K is presented in figure: 6.6. Though weak ferromagnetism has been exhibited by the sample, its remanence magnetization (Mr) and coercive field can be realised within small magnetic field. Although, all three Mn doped ZnO samples showed similar features, but, Mn-3 exhibited highest coercive field of H_c = 96.41 Oe at 300 K, while Mn-1 showed negligible amount of remanence magnetization and coercive field at 300K. The values of remanence magnetization and coercive field at 300K exhibited by Mn-3 and Mn-5 are shown in the table: 6.2.

Table 6.2: Remanence magnetization (Mr) and coercive field (Hc) exhibited by Mn-3 and Mn-5 at 300K.

Sample	M_r (emu)	H_c (Oe)
Mn-3	9.5×10^{-5}	96
Mn-5	5.6×10^{-5}	69

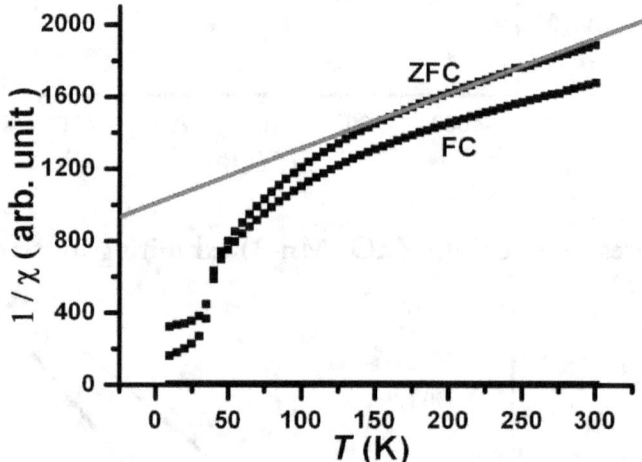

Figure 6.7: Inverse of magnetic susceptibility vs. temperature (ZFC and FC) curves of Mn-3 system.

The inverse of susceptibility vs. temperature of ZFC and FC curves is being depicted in figure: 6.7. A least squares fit of the linear portion of the curves (figure: 6.7) above 100 K gave a negative Weiss temperature (θ_w) for the sample. The value of θ_w is obtained as –225 K. The high negative values of θ_w indicate either strong antiferromagnetic or ferromagnetic interactions in these Mn-doped samples. It is to be noted that there is a sharp increase in the long magnetic moment below 50 K for all Mn doped ZnO samples. A typical *Curie–Weiss* behavior is observed above 100 K. Very large negative values of θ, along with a deviation from linearity below 50 K are in consistency with other reports [310-313]. The type (ferro-antiferro) of the magnetic interaction will depend on the Mn-Mn distances. In a random mixture of Zn and Mn ions, some Mn ions could be at a shorter distances than other, resulting in antiferomagnetic coupling;

thus, increasing the Mn content will reduce Mn-Mn distances and reinforce antiferromagnetic interaction, consistent with the theory of super exchange [314].

(ii) Co doped ZnO DMS

The magnetic measurements of the Co doped ZnO samples are done by using SQUID within temperature 4K to 300K. Zero field cooling (ZFC) and field cooling (FC) of Co-3 sample at two magnetic field 0.1 T and 100 Oe was performed to investigate the influence of magnetic moment of my specimen with variation of temperature when the specimen was exposed to two different magnetic fields. The M~T response for ZFC and FC at two different magnetic field (0.1 T and 100 Oe) is presented in figure: 6.8. From the figure it was observed that at two magnetic fields, a considerable change in the magnetic moment of the specimen occurred. For higher magnetic field (0.1 T) the magnetic moment was recorded as 14.7×10^{-4}, while at low magnetic field (100 Oe) the magnetic moment went down to 1.522×10^{-4} emu. On the other hand the inset of figure: 6.8 presents the ZFC and FC plot at 100 Oe in a close view, showing a significant difference between ZFC and FC curves. From this we can infer that as synthesized sample has a good signature of existence of ferromagnetism at room temperature. In the next step I have performed M~H response with the same sample to confirm room temperature ferromagnetism in the Co-3 sample.

To have a better understanding regarding ZFC and FC exhibited by Co-3 at 100 Oe, the M~T curve is presented in the figure: 6.9a in a wide view. The significant difference between ZFC and FC curve indicates ferromagnetic nature of the sample with Curie temperature above 300K. The figure: 6.9(b) and figure: 6.9(c) show the M~H curve for Co-3 doped ZnO sample at 10K and 300K respectively.

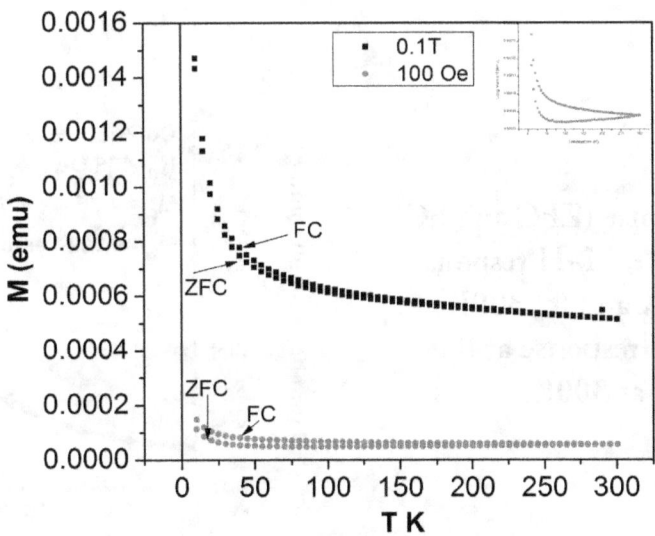

Figure 6.8: ZFC and FC response of Co-3 at 0.1 T and 100 Oe. **Inset:** exhibiting considerable difference between ZFC and FC for the sample at 100 Oe.

The Hysteresis loop exhibits ferromagnetic nature of the sample at room temperature. At 10K, the measured saturation magnetization (M_s), remanence (M_r) and coercive field (H_c) are 0.61 emu/g, 0.003 emu/g, and 104 Oe respectively. At room temperature the corresponding values of saturation magnetisation and remanence decrease to 0.0024 emu/g and 4.7E-4 emu/g respectively but the coercive field increases upto 136 Oe. An increase in coersive field at room temperature by ~ 31% was recorded for Co-3 sample. The inset in figure: 6.9b and figure: 6.9c show selective response of M~H curves at 10K and 300K respectively.

The corresponding values of remanence magnetization (M_r), saturation magnetization (M_s) and coercive field (H_c) for the sample Co-3 at 10K and 300K is presented in the table: 6.3.

In a similar work, ZHOU Shao-Min et al. [315] considered three possibilities for the origin of ferromagnetism in Co doped ZnO nanostructures. The first one is the carrier-induced ferromagnetism (RKKY or double exchange mechanism) that is often reported for the III-V semiconductors [316-318].

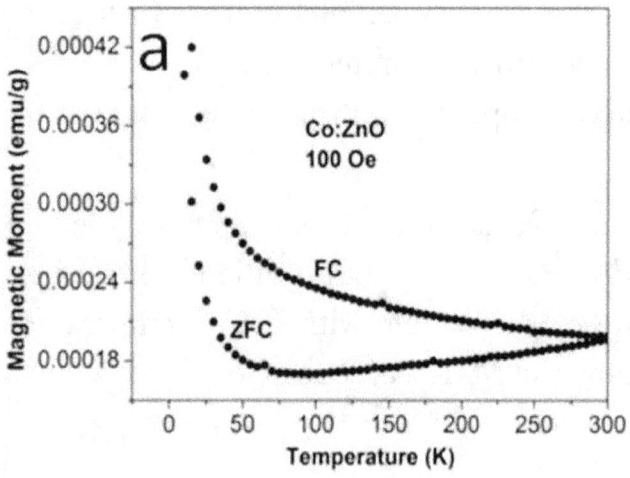
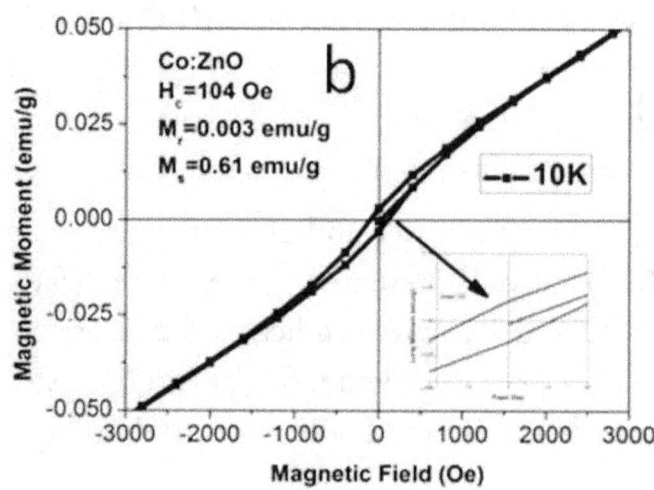

Figure 6.9: (a) M~T response (ZFC and FC at 100 Oe) of Co-3 sample, M~H response of the sample at (b) 10K and (c) 300K; inset: (b) selective M~H response at the origin at 10K, (c) at 300K.

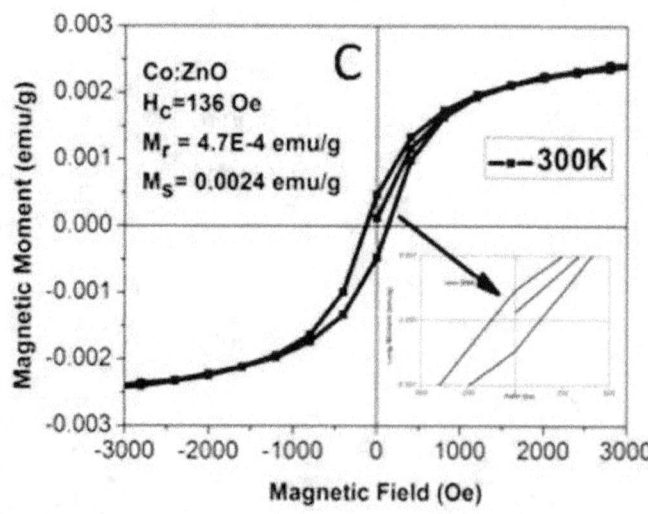

Table 6.3: Values for remanence magnetization (M_r), saturation magnetization (M_s) and coercive field (H_c) for Co-3 corresponding to 10K and 300K.

Temp.	Mr (emu)	Ms (emu)	Hc (Oe)
10K	0.003	0.61	104
300K	4.7×10^{-4}	0.0024	136

The second possibility is the presence of weak ferromagnetism due to CoO phase, whereas CoO is well known for its antiferromagnetic nature [319-321]. The third possible origin is the existence of micro Co clusters [322] in the sample. In my XRD, HRTEM, SAED and EDX study no secondary phases of Co or CoOs were detected. These observations indicate that the Co ion systematically substitute for Zn sites without changing the wurtzite structure of the samples. Thus we conclude that the observed ferromagnetism is due the presence of free carriers and localized spins. The defects produced due to Zn interstitials and O vacancies usually induce n-type characteristics.

All Co doped ZnO samples show ferromagnetic behavior at room temperature. The Co-3 doped sample shows maximum saturation magnetization 0.0024 emu/g and coersive field 136 Oe. at room temperature. The origin of the ferromagnetism is still a source of active research and may involve a carrier-induced magnetism. We have adopted a simple fabrication method which can be applied to fabricate other TM doped ZnO nanostructures for future spintronic applications.

(iii) Ni doped ZnO DMS

Till date, fewer studies have been focused on the Ni-doped ZnO system, in which diverse magnetic properties have been observed. Super paramagnetic behavior has been observed above 30K [323], while, Paramagnetism has been reported by Yin *et al* [276], which may be an indication of the occurrence of NiO or Ni precipitates. Besides, RT FM has also been realized recently in Ni-doped ZnO samples. Liu *et al* [324] have observed RT FM in Ni-doped ZnO thin films prepared by pulsed-laser deposition, which is related to the n-type carriers that ultimately arise from oxygen vacancies in the films according to the carrier-mediated mechanism. In addition, the RTFM observed in the bulk polycrystalline Ni doped ZnO samples prepared with the sol–gel technique has been explained on the basis of the impurity d-band splitting model [325]. Moreover, Mao *et al* [326] have also obtained RT ferromagnetic Ni-doped ZnO samples prepared by solid state reaction, but FM originates from the nanosized Ni clusters formed from the decomposition of NiO during calcinations.

As fabricated Ni doped ZnO sample (Ni-3) was characterized by using SQUID, initial FC and ZFC measurements indicated ferromagnetic nature of the sample at both low and room temperature which is shown in the figure: 6.10. The magnetic field during FC and ZFC measurement was kept constant at 0.1 T. It was observed that above 250 K, a slight difference exist in between the ZFC and FC curves which can be realised in a close view.

- The M~H response of Ni-3 at 10K and 300K are presented in figure: 6.11 and 6.12. The clear hysteresis loop at 10K indicates strong ferromagnetic nature of the sample at low temperature with saturation magnetization 0.1237 emu. Compared to this, M~H response at 300K exhibits weak ferromagnetism which can be visualized from the comparative curve as presented in figure: 6.13. It has been observed that the saturation magnetisation exhibited by the Ni-3 sample at room temperature cannot be realised at higher fields. It exhibits hysteresis nature below 1 T, but M~H curve is straight above 1 T.

As observed from the figure that the paramagnetism was predominant at higher magnetic fields. The establishment of the paramagnetism at higher magnetic fields also explains the no-saturation of the M-H curves at low fields.

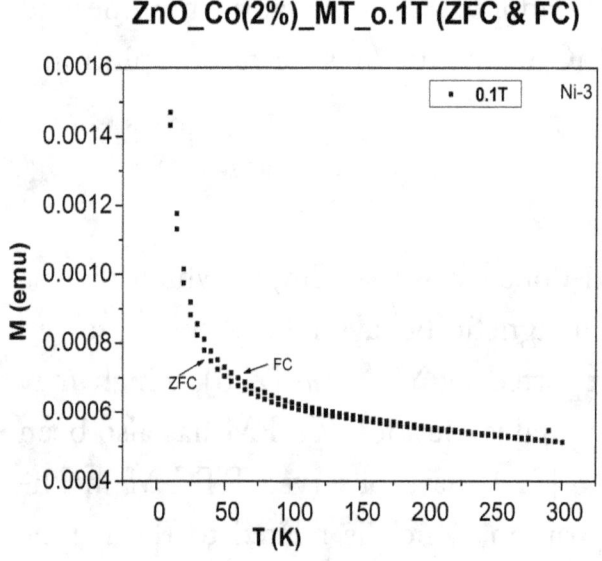

Figure 6.10: (A) M~T response (ZFC and FC at 0.1 T) of Ni-3 sample

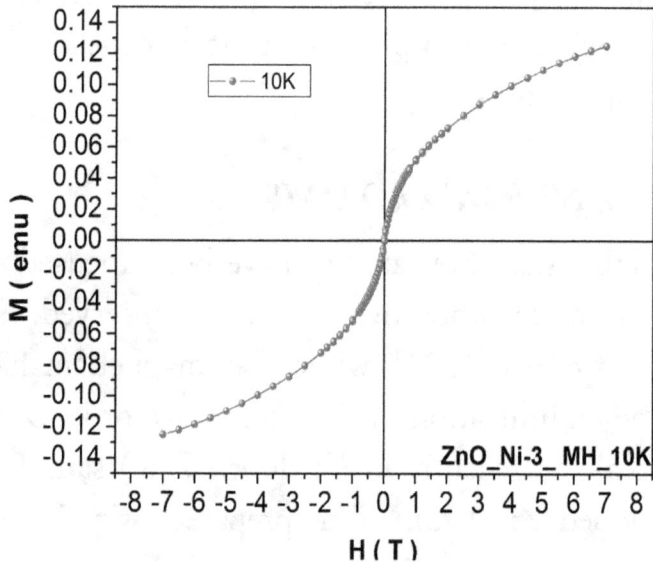

Figure 6.11: M~H response of the sample (Mn-3) at 10K.

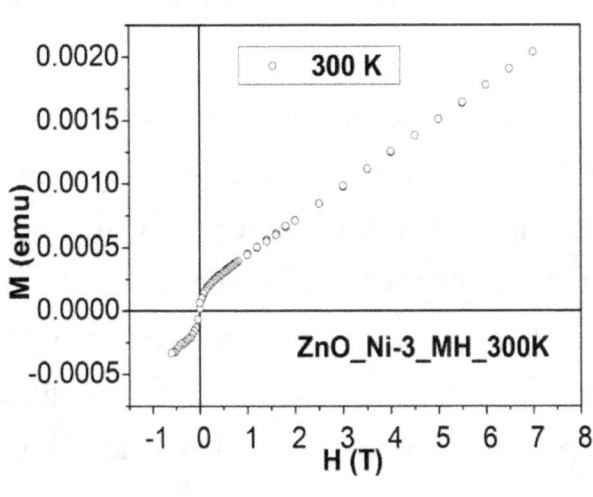

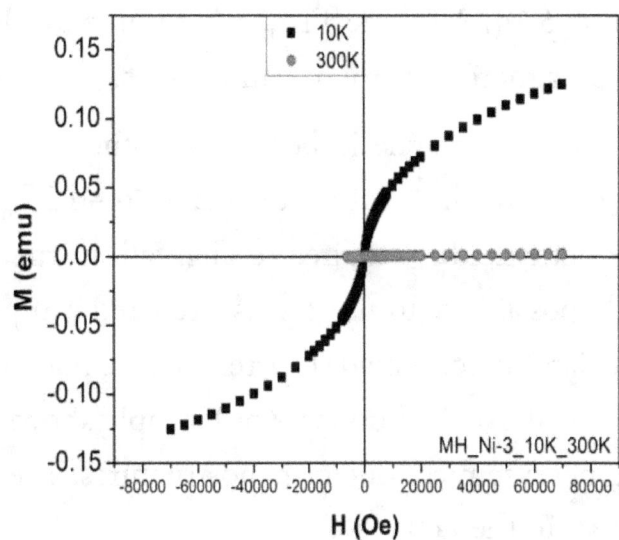

Figure 6.12: M~H response of Mn-3 at 300K.

Figure 6.13: Comparative M~H response for Ni-3 at both 10K and 300K.

The strong paramagnetic behavior at high fields can be attributed to the presence of magnetic dipoles located on the surface of nanocrystals that exhibited a minimum interaction with their neighbours inside of the crystal. Consequently, the interchange energy in those magnetic dipoles would be reduced increasing their freedom to get re-oriented. Therefore, since diminishing the crystal size will increase the crystal surface to crystal volume ratio, the population of dipoles ordered in the same direction will decrease. Thus, the sum of the total amount of dipoles oriented along the same direction will also decrease. In short, and from a magnetic ordering viewpoint, the crystal surface will usually be less magnetically ordered than the centre of the nanocrystal [327].

Table 6.4: Values for remanence magnetization (M_r), saturation magnetization (M_s) and coercive field (H_c) for Mn-3, Co-3 and Ni-3 at 300K.

	Mr (emu/gm)	Ms (emu/gm)	Hc (oe)
Mn:ZnO	2.99×10^{-4}	.0024	69
Co:ZnO	4.5×10^{-4}	.0035	136
Ni:ZnO	2.23×10^{-4}	8.53×10^{-4}	192

The dependence of magnetic moment with variation of magnetic field at room temperature for transition metal (Mn, Co, Ni) doped ZnO is presented in a single M~H plot as shown in figure: 6.14. All three TM:ZnO sample exhibited clear hysteresis loop indicating existence of ferromagnetism at room temperature.

The ferromagnetic behavior of the samples had been observed within small range of magnetic field; it was about -0.1 T to +0.1 T. So, it can be revealed that very small amount of magnetic field is required to align all magnetic moments within the nanocrystals. This opens wide possibility to use the as-prepared samples for future applications at room temperature. The data coercive field (H_c), remanence magnetization (M_r) and saturation magnetization (M_s) for the above TM doped ZnO sample shows values which are consistant with typical diluted magnetic semiconductor nanostructures. The corresponding values of M_r, M_s and H_c data are shown in the table: 6.4.

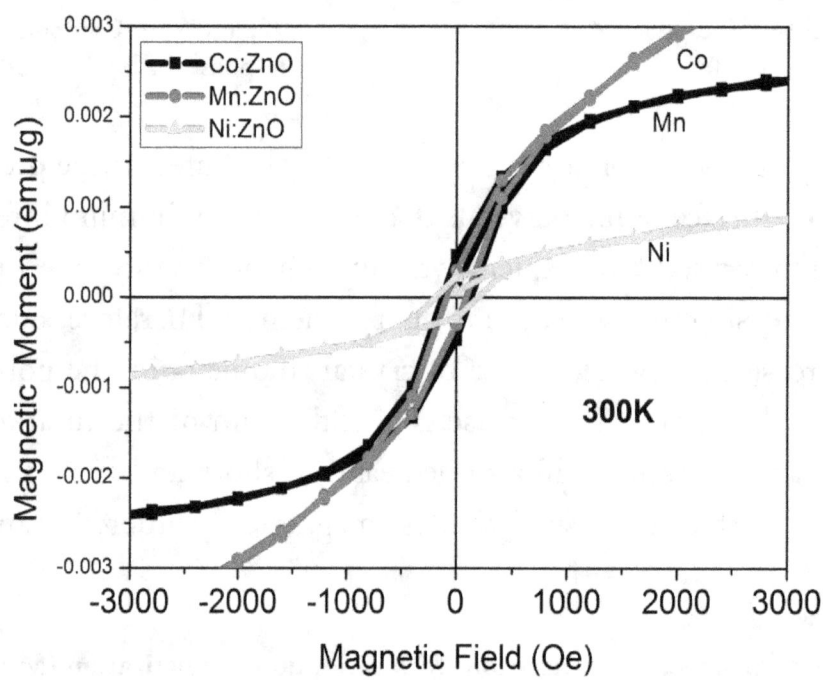

Figure 6.14: Comparative M~H response for Ni-3 at both 10K and 300K.

Apparently, the magnetic properties observed so far in Ni:ZnO strongly depend on the methods and conditions used in the preparation. The origin of RT FM observed in such a system remains a controversial subject. It is not yet clear whether the observed phenomenon is truly intrinsic or related to secondary phases such as Ni clusters [328]. However, we believe that, the weak ferromagnetic interaction as observed at room temperature may be due to intrinsic property of the nanocrystal and also 'O' vacancy in the system. It is to be noted that, we have not observed any evidence of the existence of secondary phases during EDX, XRD and TEM studies. In fact, existence of small amount of secondary phases in the sample cannot be ruled out because my characterization tools may not go beyond to detect them.

6.3 Magnetic measurement of TM doped ZnS

(i) Mn doped ZnS nanostructure

(a) MFM studies

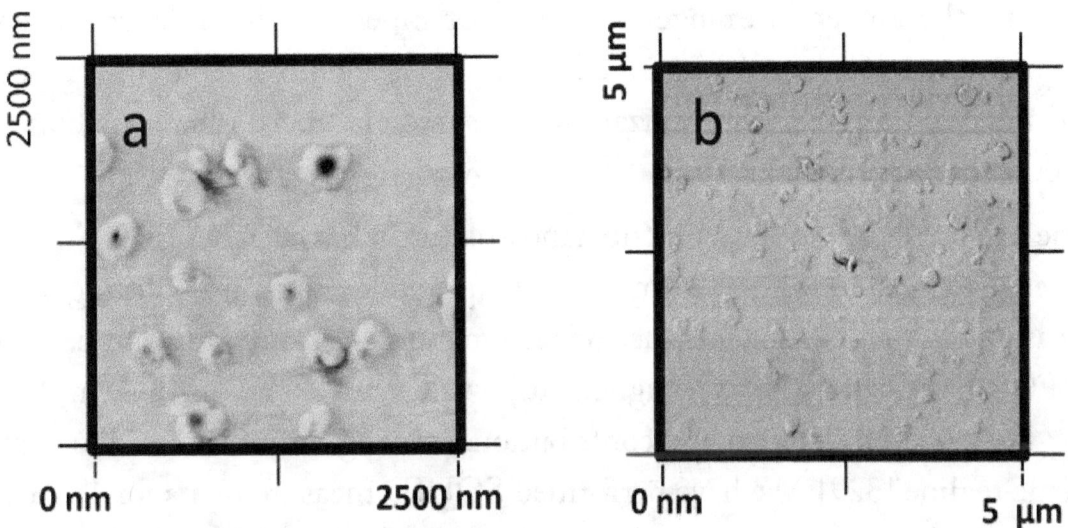

Figure 6.15: Magnetic force microscopy (MFM) image of Mn doped ZnS nanostructures exposed to the sample-C embedded in PVA. Two images with magnification surface area (a) 2500 nm X 2500 nm and (b) 5μm X 5μm are shown.

Magnetic force microscopy is a well established method to probe the micro-magnetic properties of samples with lateral resolution down to ~50 nm. The advantage of MFM is that less sample is needed, thinning or polishing of the sample is not necessary. Moreover, the technique yields information on both the structural (AFM mode) as well as the magnetic (MFM mode) aspects with regard to sample's surface. Therefore, the topology and magnetic domain structure of a sample may be correlated efficiently at the nanometer scale. Figure: 6.15a and 6.15b show MFM image of ZnS:Mn (sample C) at two different magnifications. In figure: 6.15b, nanoparticles with their clear response to magnetic field has been visualized. The average size of the magnetic cores (black spot) in figure: 6.15a is measured as ~60 nm (actual particle size) and white regions spreading over the cores are ascribed to the region of influence by the respective particles. Sharp contrast of the magnetic cores, encircling white bands is seen for many isolated particles. Therefore, it is evident that ZnS:Mn nanoparticles can respond appreciably to magnetic force and fields and MFM in this regard, can be a very good tool to exploit magnetic domains and particle-particle interactions.

With MFM studies, one could obtain visual information with regard to magnetic cores and influence zone. Investigation of Mn-incorporated diluted semiconductor nanosystems would be promising in the area of spintronics and other miniaturized magnetic & optoelectronic devices.

(b) SQUID measurements

We have investigated the magnetic properties of Mn doped ZnS samples using a SQUID magnetometer in the temperature range 4–300 K. The temperature dependence of magnetization is shown in figure: 6.16 for 3 at.% Mn doped ZnS. The figure shows a plot of zero-field-cooled (ZFC) and field-cooled (FC) magnetization measurements performed on the materials by applying 100 Oe. magnetic field.

From the M~T response of the Mn:ZnS nanostructure it has been observed FC curves which give indicates absence of ferromagnetism in the sample. Since, a distinct splitting between ZFC and FC measurement reveals the presence of a magnetic transition temperature. The difference between ZFC and FC gives the net magnetization value (ΔM = FC – ZFC) in the sample by eliminating the para and diamagnetic contributions, leaving only the contributions from the ferromagnetic regime [329]. We have performed SQUID measurements for 1at.% Mn doped ZnS sample, but observed same nature that exhibited by the 2 at.% Mn doped sample. Inset of figure: 6.16 represents the inverse susceptibility curve of ZFC plot for Mn:ZnS sample at 100 Oe. The inverse susceptibility curve is almost linear which exhibit curie Weiss behavior and indicates that the as-synthesized sample is paramagnetic in nature.

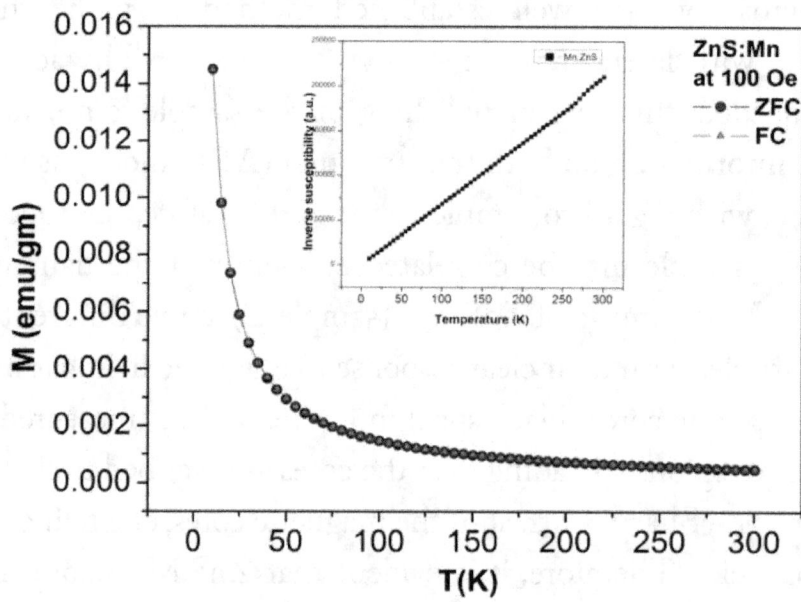

Figure 6.16: Temperature dependence of magnetization (ZFC and FC) for Mn doped ZnS at 100 Oe., inset: Inverse susceptibility curve for ZFC curve of Mn:ZnS sample.

(ii) Cr doped ZnS nanostructures

(b) MFM studies

To exploit magnetic properties, we have carried out MFM studies on the Cr doped ZnS samples. A typical MFM micrographs on a section of the ZnS:Cr sample in PVP is shown in figure: 7.17. The micrograph yields information regarding magnetic (MFM mode) aspects with regard to sample's surface. The topology and magnetic domain structure of the sample are correlated efficiently at the nanometer scale. Moreover, the technique yields information on both the structural (AFM mode) as well as the magnetic (MFM mode) aspects with regard to sample's surface. Therefore, the topology and magnetic domain structure of a sample can efficiently be correlated at the nanometer scale. Figure: 6.17(a) depicts uniform distribution of magnetically influenced nanoparticles.

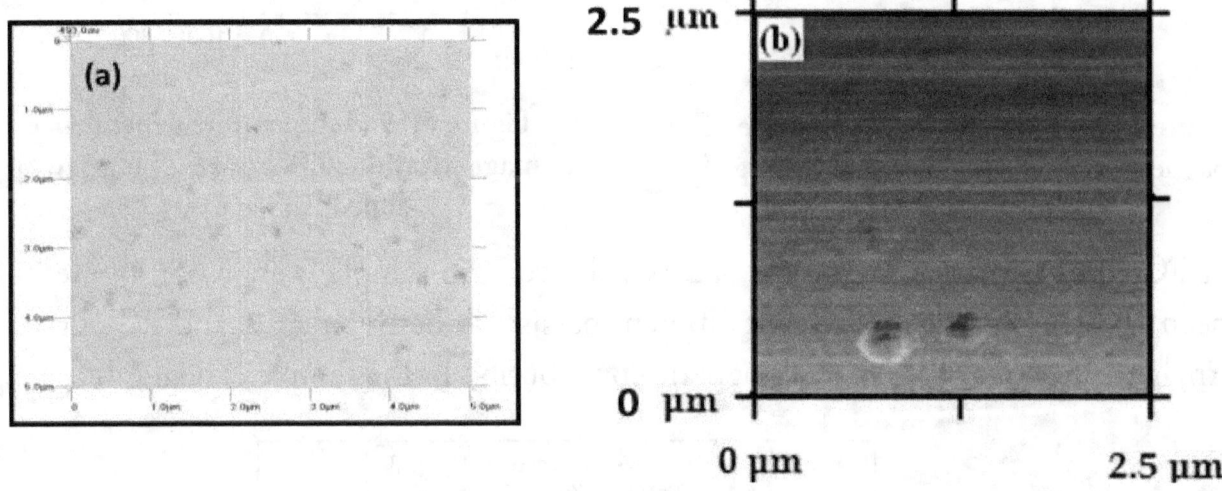

Figure 6.17: (a) Magnetic force microscopy (MFM), showing coloured image of magnetically influenced Cr:ZnS nanostructures in PVP. and **(b)** Isolatetd MFM image for Cr doped ZnS in PVA matrix.

The figure: 6.17(b) is basically a phase image of the Cr:ZnS sample embedded in PVA matrix; it shows clear response to the magnetic field. The average size of the magnetic cores (black spot) is measured as ~ 165 nm and white regions spreading over the cores are ascribed to region of influence by the respective particles. Thus, it is evident that ZnS:Cr nanoparticles can respond appreciably to magnetic force and fields and MFM in this regard, can be a very good tool to exploit magnetic domains and particle-particle interactions.

(b) SQUID measurements

For magnetic characterization by using SQUID magnetometer, we have prepared Cr:ZnS samples for two different Cr concentration, 1at.%Cr and 2at.%Cr. ZFC and FC measurements were carried out for both the samples. The dependence of magnetic moment with variation of time (ZFC plot) for 1 at%Cr at 100 Oe. magnetic field is presented in figure: 6.18. At low temperature, below 100 K, the magnetic moment increases above 30×10^{-6} emu, which is very small compared to other typical diluted magnetic semiconductors. This indicates paramagnetic nature of the sample.

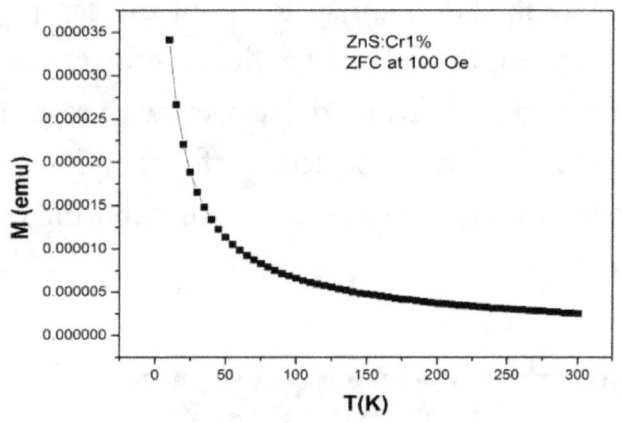

Figure 6.18: Temperature dependence of magnetization (ZFC) for 1at.%Cr doped ZnS at 100 Oe..

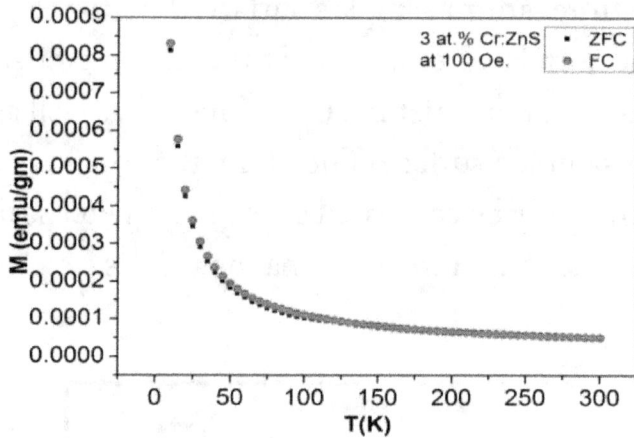

Figure 6.19: Temperature dependence of magnetization (ZFC and FC) for 3at.%Cr doped ZnS at 100 Oe..

The ZFC and FC measurements for 3at.%Cr doped ZnS sample is shown in figure: 6.19. It has been observed that no significant different persists in between ZFC and FC curves. This in turn, indicates that the net ferromagnetic contribution in the sample is negligible.

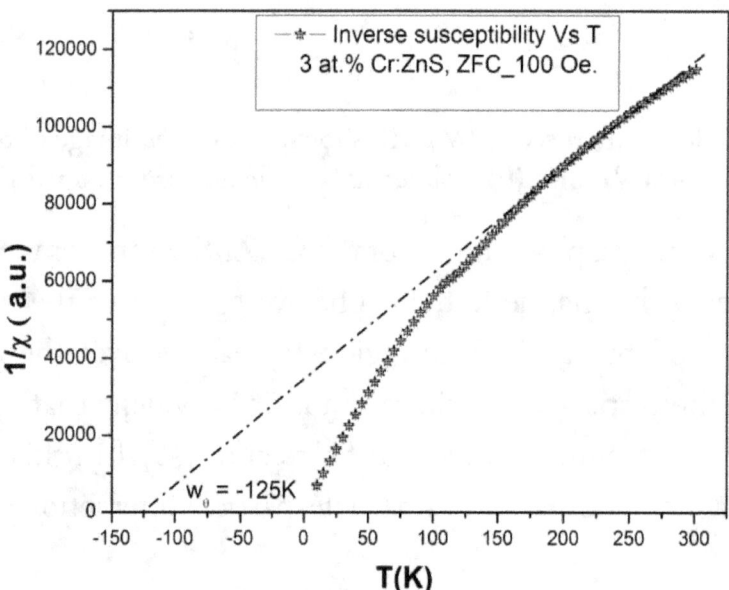

Figure 6.20: Inverse susceptibility Vs Temperature plot for ZFC of 3at.%Cr doped ZnS at 100 Oe.. The dotted line cuts negative Curie-Weiss temperature ($w\theta$) 125K.

The inverse susceptibility Vs temperature variation plot for this sample is presented in the figure: 6.20, which exhibits Curie-Weiss behaviour with negative Curie-Weiss temperature ($\omega\theta$ = - 125K). A sharp increase in magnetic moment below 50K along large negative value of Curie-Weiss temperature suggests that the dominant spin-spin interaction were of strong antiferromagnetic type.

From the above investigation on Cr:ZnS system, though 3at.%Cr doped sample exhibited noticeable magnetic moment compared to 1at.% Cr sample, yet it was far away from exhibiting room temperature ferromagnetism. It can be inferred that Cr concentration in ZnS plays an important role influence magnetic moment in the specimen. From the study, it can be speculated that due to the absence of ferromagnetism in the sample, there is a less possibility to apply them in practical applications. Since, spintronics or any other electronic appliances room temperature ferromagnetism is the most important factor.

6.4 Significant observations

- M~H responses of bare ZnO exhibited typical diamagnetic nature.
- In the M~T measurements at 300K, all TM (Co, Mn, Ni) doped ZnO nanostructures displayed significant difference between FC and ZFC curves, which indicated the existence of ferromagnetism in the samples where Curie temperature is above 300K.
- M~H responses of TM: ZnO samples were carried out at room temperature for all TM doped ZnO samples exhibited typical hysteresis loop, which evidenced the existence of ferromagnetic coupling in the samples.
- We have investigated the influence of Mn content in the magnetic properties of Mn:ZnO system. As per observation, at low Mn^{2+} ion concentration (Mn-1) magnetic responses (H_c, M_r, M_s) were minimum, at medium Mn^{2+} ion content (Mn-3) these became maximum, again at increased Mn^{2+} ion content (Mn-5) these responses were lowered. For example, in the M~H response at room temperature, Mn-1 showed negligible but evident coercive field, in turn, the respective values for Mn-3 and Mn-5 were recorded as 96 Oe. and 69 Oe. respectively. Thus, it can be reasonably inferred that TM ion concentration in diluted magnetic semiconductors plays a vital to influence magnetic properties in the specimen.
- In the susceptibility curve of Mn-3, a typical *Curie–Weiss* behaviour was observed above 100 K along with large negative values of *Curie–Weiss* temperature (θ_w = - 225K) along with a deviation from linearity below 50 K, which suggests that there exist either ferromagnetic or antiferromagnetic couplings in the specimen.

- All Co doped ZnO samples show ferromagnetic behaviour at room temperature. Compared to Mn:ZnO and Ni:ZnO systems, a significantly large difference between ZFC and FC curves has been observed in Co:ZnO system, which indicated existence of strong ferromagnetic coupling in Co:ZnO system. A linear increase in magnetic moment with increasing magnetic field was displayed by Co:ZnO system. The M~H response for Co-3 studied at 10K as well as 300K exhibited clear hysteresis loops indicating ferromagnetic nature of the sample at both temperatures. Compared to 10K measurements, the corresponding values of saturation magnetisation and remanence at room temperature have lowered, but the coercive field has increased by ~31%.

- The M~H responses of Ni:ZnO system showed strong ferromagnetic coupling at 10k, but comparatively weak ferromagnetism at room temperature. For Ni-3, saturation magnetisation at room temperature was obtained at low magnetic field, but at higher field, no saturation was obtained. The strong paramagnetic behaviour at high fields can be attributed to the presence of magnetic dipoles located on the surface of nanocrystals that exhibited a minimum interaction with their neighbours inside of the crystal.

- In the MFM study, uniform distribution of magnetic domains for TM(Cr, Mn) doped ZnS system were detected.

- In contrast to TM:ZnO system, all TM(Mn, Cr) doped ZnS sample exhibited paramagnetic behaviour. M~T response in case of Mn:ZnS system is more significant than Cr:ZnS. In the M~T response of Cr:ZnS system, 2at.% Cr doped sample displayed slight higher magnetic moment compared to 1at.% Cr from which we can infer that TM ion content in ZnS can play a role to influence magnetic moment in the specimen.

- The origin of the ferromagnetism is still a source of active research and may involve a carrier-induced magnetism. In my XRD, HRTEM, SAED and EDX study no secondary phases due to TM incorporation were detected. These observations indicate that the TM ion systematically substitute for Zn sites without changing the wurtzite structure of the samples. It suggests that the observed ferromagnetism in TM:ZnO may be purely intrinsic and due to the presence of free carriers and localised spins. The defects produced due to Zn interstitials and O vacancies usually induce n-type characteristics.

- Finally, we conclude that all TM(Mn, Co, Ni) doped ZnO nanostructures exhibited room temperature ferromagnetism. From the comparative M~H response of all TM doped sample it was found that Co:ZnO system exhibited largest value of remanence magnetization (4.5×10^{-4} emu/gm), saturation magnetization (3.5×10^{-3} emu/gm) along with coercive field 136 Oe. From this we can reasonably infer that Co-3 can be good candidate for device making purpose. This significant result can be applied to luminescence / spintronic applications.

6.5 Possible applications in spintronic devices

Spintronc has a vast area for research. A good number of people are engaged to explore spintronic devices for next generation technology. Along with modern sophisticated tools adequate requirements are needed to achieve the goal towards practical spintronic devices. My noble approach was to fabricate good quality DMS sample so that it could be utilized for future spintronic applications.

In 1999, Y. Ohno [330] and his group performed an experiment on electrical spin injection in a ferromagnetic semiconductor heterostructure. They have reported the fabrication of all-semiconductor, light-emitting spintronic devices using III-V heterostructures based on gallium arsenide. Electrical spin injection into a nonmagnetic semiconductor was achieved (in zero magnetic field) using a p-type ferromagnetic semiconductor as the spin polarizer. Spin polarization of the injected holes was determined directly from the polarization of the emitted electroluminescence following the recombination of the holes with the injected (unpolarized) electrons. They performed their experiment below room temperature, since Curie temperature of III-V semiconductor was within 40^0-90^0 C. In their Spin-injection set up, the experiment was performed under 0.1 T magnetic field with 20 nm spacer. This work encouraged other researchers to develop Spin-LED.

The most promising candidates for high-temperature semiconductor spin aligners appear to be the diluted magnetic alloys Ga–M–N and Zn–M–O (where, M = V, Cr, Mn, Fe, Co and Ni). Mean-field calculations performed by Dietl et al [335] predicted that these materials could exhibit ferromagnetism at and above room temperature upon doping with transition metal elements on the order of 5 at.% material. High-temperature ferromagnetism in these materials stems partly from their strong p–d hybridization owing to their small interatomic spacing and small spin–orbit coupling.

Another promising approach for efficient electrical spin injection is through the use of a diluted magnetic semiconductor (DMS) spin aligner in which the conductivity mismatch can be greatly reduced. In II–Mn–VI paramagnetic semiconductors such as CdMnTe and ZnMnSe, the sp–d exchange interaction between itinerant electron spins and localized Mn^{2+} ions results in a large Zeeman splitting of the conduction band states under the presence of an applied magnetic field. The Zeeman splitting is given by $E = g_*\mu_B H$ where g_* is the effective electron g-factor. With no external magnetic field, the exchange fields due to the Mn^{2+} ions are randomly oriented and cancel one another. Application of a magnetic field creates a net magnetization of the Mn spins and a non-zero Zeeman splitting. When an unpolarized current supplied from a non-magnetic metal contact is driven through the DMS layer, the injected electrons quickly scatter into the energetically lower spin sub-band and become spin-polarized within a picosecond along the magnetic field direction. These spin-polarized carriers may then travel by drift or diffusion into an adjacent non-magnetic semiconductor.

These and similar predications encouraged intensive experimental activity aimed at developing transition metal doped ZnO for spintronic applications.

My present study suggests that out of all TM doped ZnS and ZnO samples few of them can be considerably employed for next generation technology. Room temperature ferromagnetism in DMS was a big challenge, we have achieved this goal by fabricating TM doped ZnO samples adopting very simple inexpensive chemical method. It opens a large possibility of research in this emerging area. To fabricate nanostructures for research purpose, all laboratories cannot provide sophisticated tools like MBE, in turn; chemical approach can be available everywhere for large scale fabrication. By adopting this simple method we have fabricated al TM doped ZnO nanostructures with sizes below 20 nm, which is very much promising for low dimensional spintronic devices. As per my investigations, the saturation magnetization for all TM doped ZnO samples were obtained at low magnetic field (below 1 T). This is encouraging, because my samples could be controlled by applying adequate magnetic field.

There are a number of requirement for achieving practical spintronics devices. The most fundamental requirements are:

- The capability to transport the carriers with high transmission efficiency within the host semiconductor or conducting oxide;
- The capability to detect or collect the spin-polarized carriers and to be capable of controlling transport through external means such as biasing of a gate contact on a transistor structure. These type aspects of spin injection, spin-dependent transport, manipulation and detection form the basis of current research and future technology.

6.5.1 Proposed Scheme for spin aligner in Spin LED

The optical polarization state may be characterized through many techniques, but most rely on a quarter-wave retarder and linear polarizer to decompose the electroluminescence into its right- and left-circularly polarized components. A typical experimental setup is shown in figure:

6.21. A lens is used to collimate the emission from the spin-polarized light sources, which then passes through the retarder–polarizer optics. Calcite polarizers transmit satisfactorily from 0.3 to 2.3μm, which is more than adequate for all spin-polarized light sources considered to date. Quarter-wave retarders, on the other hand, are designed for a particular wavelength and must be matched to the wavelength of the light source. This is particularly troublesome for analysis of spin-polarized light sources containing wide band gap materials in the active region due to the lack of suitable retarders. For incoherent light sources emitting over a broad spectra range, an achromatic quarter-wave plate centred at the peak emission wavelength may be desirable. After passing through the retarder–polarizer combination, a collection lens is used to focus the light onto a spectrometer or photodetector. From the Mueller matrices for a rotating quarter-wave plate and linear polarizer, it may be shown that the intensity of the emission emerging from the retarder-polarizer combination is given by

$$I(\theta, \phi) = 1/2[S_0 + S_1/2 + S_3 \sin(2\theta) + (S_1/2)\cos(4\theta) + (S_2/2)\sin(4\theta)] \qquad \rightarrow 1$$

where S_0, S_1, S_2 and S_3 are the familiar Stokes parameters.

Degree of Circular polarization will be measured as

$$\pi_{CP} = S_3/S_0 \qquad \rightarrow 2$$

It is the ratio of two stokes parameters.

and $\theta = \omega t$, is the angle of rotation of the quarter-wave plate fast axis.

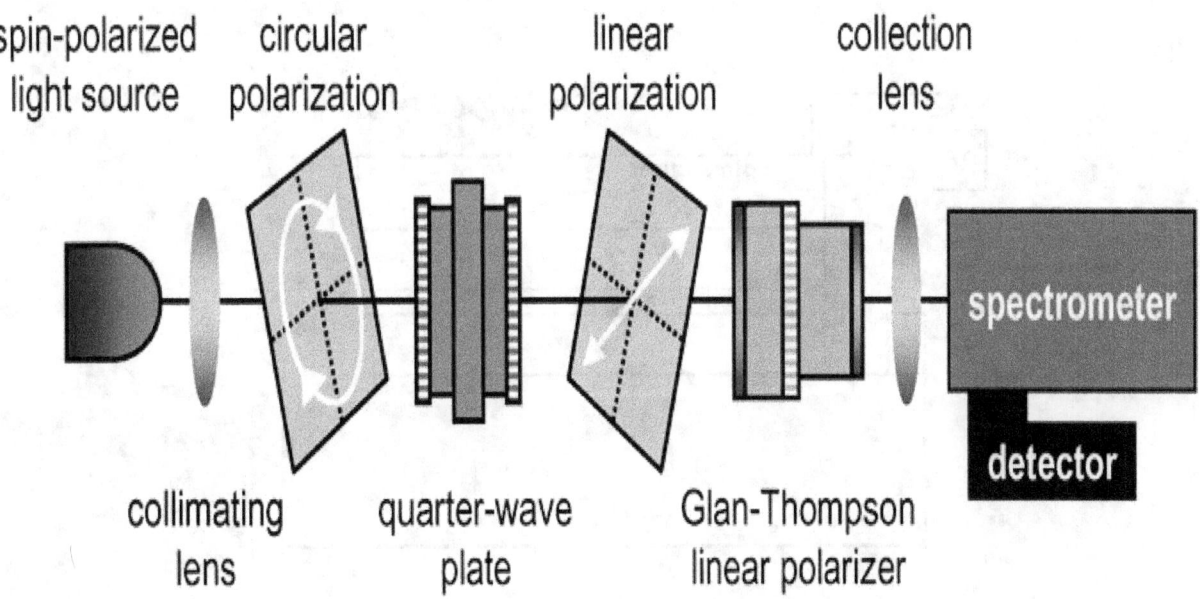

spin-polarized   circular   linear   collection
light source   polarization   polarization   lens

spectrometer

detector

collimating   quarter-wave   Glan-Thompson
lens   plate   linear polarizer

Figure 6.21. Typical experimental setup for characterization of spin-polarized light sources [Ref:331].

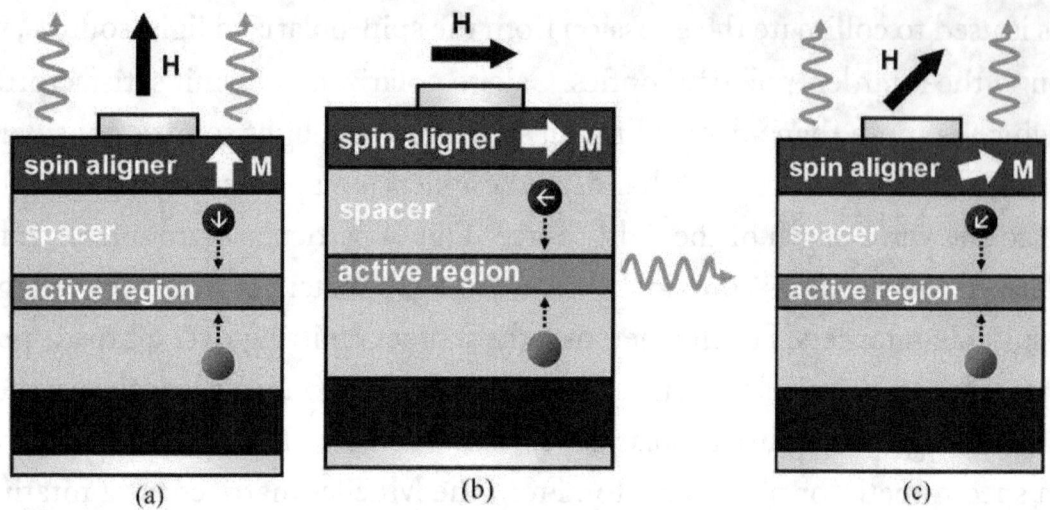

Figure 6.22. Schematic representation of a spin-LED under the (*a*) Faraday, (*b*) quasi-Voigt and (*c*) oblique Hanle effect geometries. [Ref: 331]

The goal of most experiments involving spin-polarized lights sources is to relate the device's optical polarization to the spin polarization of carriers injected from a magnetic spin-aligner layer into a non-magnetic semiconductor. Three measurement geometries are typically employed for the characterization of spin-polarized light sources [331]: (a) Faraday, (b) Quasi-Voigt and (c) Oblique Hanle effect geometries [figure: 6.22].

Spin-LED measurements are most commonly performed in the Faraday geometry since the selection rules are best understood in this configuration, allowing a fairly direct conversion between spin and optical polarizations. A typical experimental setup under the Faraday geometry as from Ref: 331 is shown in figure: 6.21.

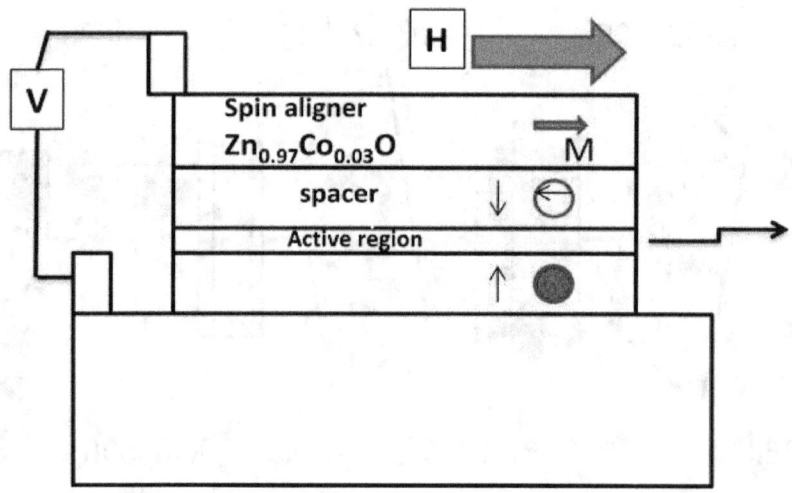

Figure 6.23: Proposed Scheme of Prototype Spin-LED.

After merging my idea to develop a spintronic device along with the discussion above we find maximum possibility to utilise my as-fabricated Co:ZnO sample as a spin aligner in Spin-LED model. In most of the Spin-LED sources, GaMnAs, InGaAs and ZnMnS DMSs were utilized as spin aligners. As per my present investigations on II-VI based transition metal doped diluted magnetic semiconductors, we have observed that TM(Co, Mn, Ni) doped ZnO exhibited few desirable basic properties as predicted by Dietl et al. [332]. These samples with low dimensional size (less than 20 nm) displayed room temperature ferromagnetism and showed the possibility to operate within 0.1T magnetic field. Out of all samples, Co:ZnO system, doping with 3at.%Co exhibited significant magnetic properties which are similar to the typical diluted magnetic semiconductors. From these observations, we can reasonably infer that Co-3 sample could be promising candidate as a spin aligner for efficient spin injection.

A typical prototype Spin-LED utilizing $Co_{0.03}Zn_{0.97}O$ as spin-aligner as proposed by us is shown in figure: 6.23. As we have discussed in chapter-1 that the most straightforward approach to design a Spin-LED would be to implant Mn into the top contact p-GaN layer of the standard GaN/InGaN LED. Here, we propose to use $Co_{0.03}Zn_{0.97}O$ as spin aligner, since this particular sample have achieved few basic requirements as discussed earlier. In the figure: 6.22 'H' is the external magnetic field (~ 1T). The LED source is set in the quasi-Voigt geometry. In presence of the external magnetic field the spin of the spin aligner will be align parallel to the magnetic field. This could be possible for the material $Co_{0.03}Zn_{0.97}O$, since it exhibited RTFM during investigations. In turn, randomly oriented spins of the spacer will be antiparellel to the magnetic moment of the spin aligner which will make it possible to inject spin polarised light. The source can be used in the set up as shown in figure: 6.21 for measuring degree of circular polarization.

Though, I have suggested a scheme, I do not claim that these are the sufficient requirements to get circularly polarized light at the collector end. For practical visualization of the complete experimental set up more parameters have to be explored along with adequate measuring tools.

In fact, we believe that to design a 'spintronic device' is not an easy task.

6.6 Possible applications in Luminescent devices

ZnO and ZnS are the most studied wide band gap semiconductor, they have tremendous potential in luminescent device applications due to their properties like: Size dependant emission, broad excitation range, high quantum yield, etc..

a. In my study it has been observed that both undoped and organic host (PVA/PVP) coated ZnS exhibited intense blue emission. Usually, the blue emission exhibited by

ZnS is due to 'S' vacancy which can quinch out the band-edge and other defect related emissions. There is a possibility to tune the luminescence response to its requirement by controlling the amount of sulphur at the time of synthesis.

b.   Colour tuneable application: Typical PL spectra were observed for TM doped ZnS. The Mn doped ZnS exhibited orange-yellow emission at ~580 nm which opens wide possibility to control the luminescence intensity by controlling the doping concentration of the transition metalby controlling the aspect ratio of ZnO:TM nanorods the luminescence can be tuned,

c.   It was predicted that nanorods with smaller length is useful for Oxygen gas sensor, while nanorods with larger dimension is more useful in UV optical sensor. Thus, by controlling the synthesis protocol the length of the nanorods can be controlled which in turn will give the opportunity to apply them in desired Luminescent device applications.

6.6.1 Possible application in design of biological marker

Nanostructured zinc oxide (ZnO) thin films are showing an increasing potential as sensing components in electronic nose instruments. As described in [333–335], these materials have been successfully applied in the detections of volatile organic compounds particularly associated to markers of meat spoilage. With certain markers such as ethanol, the nanostructured ZnO thin films have shown detection levels in the ppb levels, thus outperforming traditional metal oxide semiconductors based on SnO_2.

In a recent paper Martin Längkvist et al. [336] proposed a fast sensor based on ZnO nanostructured thin film which can be used as fast classification of meat spoilage marker. The application area that they have considering is food safety and in particular they aimed at developing an instrument that can be used in situ for rapid identification of meat spoilage.

Mn-doped ZnO is an n-type semiconducting material. When it is exposed to the atmosphere, the oxygen molecules react with its surface and capture electrons from its conduction band. This in turn leads to a decrease in the electron concentration and, hence, increases the surface resistance until equilibrium. The stabilized surface resistance forms the baseline for the sensing studies. When the reducing vapours like ethanol or TMA are presented to the sensing element, the vapour reacts with surface-adsorbed oxygen species and increases the electrons concentration on the surface. As a result, the surface resistance decreases from the stabilized baseline and attains saturation. This change in surface resistance has a strong correlation with the concentration of ethanol/TMA in dry air atmospheric conditions [335].

In the past two decades, the awareness about food safety, particularly with respect to specific pathogenic bacteria, has increased. This is especially true in the case of meat and fish, where microbial spoilage can be dangerous for humans, and where there is a clear requirement for a rapid and accurate detection system. Traditionally, fish and meat quality is assessed by examining the structure of the food (texture, tenderness, flavor, juiciness, color), or by detecting the microorganism and its count, or by detecting the gases generated by these microorganisms. A number of techniques have been used to examine the quality of the meat, namely instrumental mechanical methods, the ultrasound technique, as well as optical spectroscopy [337,338].

In my study we have observed that the PL spectra for both bare and TM doped samples exhibited different characteristics. Intense luminescent intensities have been exhibited by both bare ZnO (figure: 6.24A) and TM doped ZnO (figure: 6.24B). Compared to doped samples, bare ZnO displayed maximum luminescence in the UV region (figure: 6.24C). In case of TM doped samples the UV intensity gets lowered along with additional peaks in the visible reason. It has already been explained in chapter-4 that the effect is attributed due to radiative recombination of d electron transfer of transition metals. This property can be utilised for colour tunable effect for sensing purpose. We have little bit extended my study to observe PL intensity of undoped ZnO nanostructures in bacterial environment.

We have selected a particular bacteria "**staphylococcus**" which is responsible mostly for different kinds of skin diseases. My study revealed that in bacterial environment PL intensity of ZnO nanostructures exhibited wide spectrum in the visible region.

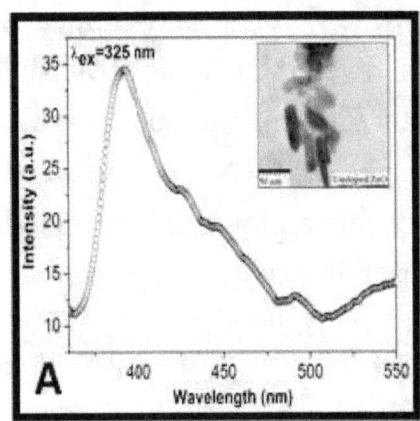 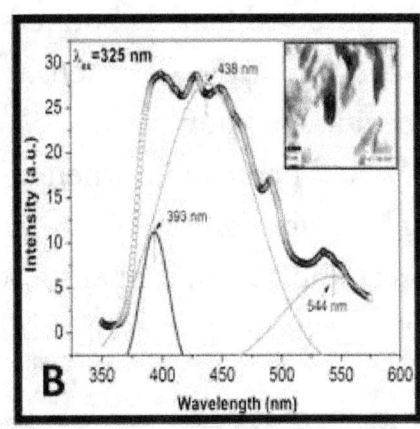 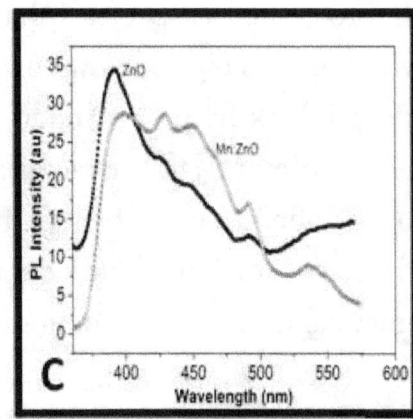

Figure 6.24: Change of PL intensity of ZnO nanorods due to Mn doping, (A): PL spectra of bare ZnO nanorods, (B) PL spectra for MN:ZnO nanorods and (C) Comparative spectra for both bare and Mn doped ZnO nanorods.

This property is remarkable, since in the visible region we can perfectly apply this sample as a biological marker to detect these particular bacteria. Initially, we have studied the PL spectra of bare ZnO nanostructure in a medium "Mueler Hintion Agar".

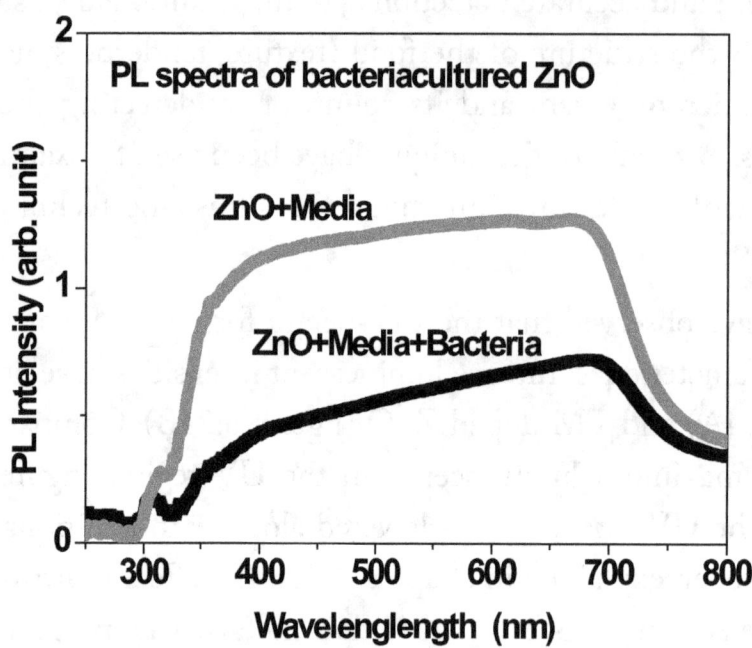

Figure 6.25: PL spectra of bare ZnO NSs with media and Staphylococcus bacteria assisted environment.

It is a microbiological growth medium that is commonly used for antibiotic susceptibility testing. The PL spectra of bare ZnO nanostructures with "Mueler Hintion Agar" medium alone and with the staphylococcus bacteria cultured environment is presented in figure: 6.25.

From the PL spectra, it has been observed that for both cases a wide visible spectra appears, when bacteria is added with ZnO and media, the PL intensity gets lowered by about 50%. The sharp UV peak of ZnO changes to wide spectrum within 350 nm to 750 nm. This suggest that more different PL characteristics would be possible when TM doped ZnO NSs are utilized for the above study. The significant behaviour of PL spectra in the visible region can be utilized to design efficient biological marker. Apparently, it opens a wide possibility but for complete practical design of a biological marker extensive measures should be considered.

6.7 Limitations

a. Time-gap

Time gap between the synthesis and characterization is the most important factor in nanoscience research. Besides our own laboratory, we had to investigate our samples at three other places. For, AFM and MFM the samples were sent to Romanioa, SQUID investigations were done at Indore and HRTEM studies were carried out at Shillong, for which there was always a time-gap between the synthesis and analysis. Thus, investigations were performed under this limiting condition for which we may not get the expected result.

b. Reproduction of samples

For proper investigations, same sample should be exposed to all characterization tools. Scientific characterization tools at different places may demand reproduction of the same sample which may bring deviations from the correlations among different investigations. This is another limitation which needs to be overcome. In fact, steps have already been initiated in this direction.

6.8 Future directions

a. Development of other binary semiconductor nanostructures:

During the study, I was restricted to ZnO and ZnS from II-VI groups. The study can be extended to other binary semiconductors. For example study can be extended to ZnSe, ZnTe, CdTe, CdSe etc.; since, these candidates are also promising in the nanoscale resigm.

b. Develpoment of other TM doped II-VI based DMS:

In my study Mn, Ni and Co were considered to dope ZnO, while Cr and Mn were considered for ZnS. This study can be extended to develop other TM doped II-VI base DMS. Few reports are available on the study of RTFM with Cu, Fe doped II-VI based DMS. Along with these two TMs, one can consider other transition metals like V, Ti, Mo etc. for extensive study in this direction.

c. Possibility of developing rare earth metal doped II-VI based DMS:

The contributions of rare earth metals in developing fluorescent lamps, lasers, fiber optics, magnetoresistive alloys, metal-halide lamps are excellent. Yttrium (Y) is used in TV red phosphor, Neodymium (Nd) and Gadolinium (Gd) are applied as rare earth magnets. There is a large possibility of extending this study by introducing rare earth metals in ZnS and ZnO semiconductor hosts to explore optical as well as magnetic properties.

d. Biological sensors:

It has been proved that fluorescent materials are the best candidates to replace organic dyes. Thus utilizing the size dependant UV colour sensing property of nanostructures biological sensing /imaging device or marker can be developed.

e. Post annealing:

During the study, we have not considered post synthesis annealing. To exploit magnetic properties and also to find origin of RTFM in the TM:ZnO samples it is necessary to anneal the samples up to at least 700° C. This is left for future study.

f. Study of transport phenomena:

Study of transport phenomena in the as-synthesised samples is quite necessary for practical realization of a spintronic device. The study of transport phenomena should through more light on spintronic based devices/activities.

g. Use of more characterization tools:

In the study, we were limited to few characterization tools due to unavailability, time factor and many other reasons. To exploit complete physical properties of the samples, this study may be extended in future by utilizing more and more modern tools.

h. TM co-doping:

Researchers have already investigated many interesting properties due to co-doping of TM in ZnO. We have investigated for individual TM doping cases. There is a large possibility to extend the study with co-doping of different TMs. For example, study can be extended with Mn and Co co- doping, since in my investigation both the TM have exhibited RTFM, so their co-doping in ZnO lattice host may exhibit attractive property. Similarly attempt should be taken to include other TMs for co-doping.

Appendix

Table 1: Various spectral regions.

Region	Wavelength (nm)
Far ultraviolet	10-200
Near ultraviolet	200-380
Visible	380-780
Near infrared	780-3000
Middle infrared	3000-30,000
Far infrared	30,000-300,000
Microwave	300,000-1,000,000,000

(Set by the Joint Committee on Nomenclature in Applied Spectroscopy)

Table 2: Relationship between light absorption and colour.

Color absorbed	Color observed	Absorbed radiation (nm)
Violet	Yellow-green	400-435
Blue	Yellow	435-480
Green-blue	Orange	480-490
Blue-green	Red	490-500
Green	Purple	500-560
Yellow-green	Violet	560-580
Yellow	Blue	580-595
Orange	Green-blue	595-605
Red	Blue-green	605-750

Table 3: Properties of PVA

Physical properties	
Glass transition temperature (^{0}K)	343
Melting temperature (^{0}K)	483
Refractive index	1.55
Specific gravity	1.55
Specific heat (J/gm K)	1.66
Thermal conductivity (W m^{-1} K^{-1})	2.0
Molar mass of single structure unit (g)	58.2
Dielectric constant	2.0

Table 4: Properties of PVP.

Molecular formula	$(C_6H_9NO)_n$
Molar mass	2.500 – 2.5000.000 g·mol^{-1}
Appearance	white to light yellow, hygroscopic, amorphous powder
Density	1,2 g/cm^3
Melting point	110 – 180 °C (glass temperature)

Table 5: HRTEM specifications at SAIF, NEHU, Shillong.

Resolution:	1.9Å to 1.4Å
Accelerating Voltage:	60-200 KV in 50 V steps.
Tilt:	±25°
Magnification:	With standard specimen gives x50 to 1,500,000.
High Resolution CCD Camera:	2.672 x 2.672 K

Paper Presented in Seminar/Conference by the Author

1. *Studies of ZnS:Mn nanoparticles embedded in dielectric polymer host,* CM Days -2008, Visva Bharati, Shantiniketan, 29-31 August-2008.

2. *ZnS:Cr Nanostructures building fractals and their properties,* **International conference**, ICANN-09, IIT, Guwahati, 2009

3. *Chromium doped ZnS nanostructures: Structural and Optical characteristics,* **International conference**, ICTOPON-2009, Allahabad, 2009.

4. *Structural characteristics of Cr:ZnS nanostructures embedded in dielectric polymer hosts,* CM-Days, Tripura, 2010.

5. *Elongated nanostructures as building blocks ZnS:Cr fractal patterns,* 55th Annual Technical Session, Assam Science Society,15th Feb'2010, Guwahati University.

6. *Tailoring the Structural and spectroscopic properties of Mn doped ZnO nanorods, **International conference** on Hybrid material,* Strasberg, France, 10-16th March, 2011.

7. *Growth and optical properties of Chromium doped ZnS nanostructures,* National seminar on Laser and Optical science, Kanoi college, 10-13 Oct., 2011.

8. Structural and spectroscopic investigations of Mn-doped ZnO nanorods, National seminar, NCSN, Tezpur University, 2011.

9. *Optical Study of ZnO nanorods fabricated at room temperature and cultured with staphylococcus bacteria,* UGC sponsored National seminar on Quantum mechanics, Laser and their application, Sonari College, 2012.

10. *Prospects and future implications of Nanotechnology in some important issues closely related with environment,* UGC sponsored International seminar, Teok College, Jorhat, 21-23 August, 2014.

11. *Exploring sustainable development in aid of Nanotechnology,* UGC sponsored National seminar, Furkating College, Golaghat, 29-30th August, 2014.

12. *Green nanotechnology for sustainable development,* UGC sponsored National seminar, NN Saikia College, Titabar, 29-30th September, 2014

13. *Impacts of Semester system: A comparative case study with the Annual system,* UGC sponsored national seminar, naharkatia College,15-16th October-2014.

14. *Fabrication of ZnO and Mn:ZnO nanostructures and thier possible applications in design of biological marker*, The 102nd Indian Science Congress, University of Mumbai, 3-7th January, 2015.

15. *Influence of Mn concentration on the change of lattice parameter of nanorods*, Hybrid photonics 2015, Tezpur University, 24-26th February, 2015.

16. *A study on the role of nanotechnologyin drug discovery*, **International conference**, Dibrugarh University, 11-13th March, 2015.

17. *Development of ZnO nanostructure based UV protected cosmetics*, UGC sponsored National seminar, Namrup College, Dibrugarh, 29-30th June, 2016.

Papers/Chapters Published by the Author

1. *D P Gogoi, U Das, G A Ahmed, D Mohanta, A Choudhury and G A Stanciu, Chromium Doped ZnS Nanostructures: Structural and Optical Characteristics* American Institute of Physics, (ICTOPON-2009), 978-0-7354-0684-1, 2009, pp-*502-507*.

2. D P Gogoi, G A Ahmed, D Mohanta, A Choudhury and G A Stanciu, *Structural and optical properties of Mn doped ZnS semiconductor nanostructures,* Indian J. Phys., ISSN-0973-1458, 2010, 84 (10), pp- 1357-1363.

3. D P Gogoi, U Das, G A Ahmed, D Mohanta, A Choudhury and G A Stanciu, *ZnS:Cr Nanostructures Building Fractals and Their Properties,* American Institute of Physics, (ICANN-2009), 978-0-7354-0825-8, 2010, pp-26-31.

4. D.P. Gogoi et al., Growth and optical properties of Chromium doped ZnS nanostructures *"International journal of Laser and Optical Science"* 2010 D.P. Gogoi, Prof. A. Choudhury, Issues and impacts of nanotechnology in relation to the environment, *Proceeding, national seminar, Sustainable management and conservation of environment,* 3-4 June, 2010, pp.86-95

5. Durga Prasad Gogoi, Gazi A. Ahmed and Amarjyoti Choudhury, *Structural and Spectroscopic Investigations of Mn doped ZnO Nanorods,* International Journal of Nanotechnology and Applications, ISSN 0973-631X Volume 5, Number 4 (2011), pp. 433-441.

6. Durga Prasad Gogoi, A study on the role of nanotechnology and population explotion of India, *Population Education, Edited by Dr. A Saikia et. al., ISBN-978-81-313-1223-0, 2011, pp-369-376.*

7. Durga Prasad Gogoi, *Some issues related with the tour from traditional Phytotherapy to Nanomedicine,* Traditional Phytotherapy, Edited by Dr. Jitu Buragohain, ISBN-978-81-7035-821-3, 2013, pp-110-115.

8. Dr. Durga Prasad Gogoi, *Exploring sustainable development in aid of nanotechnology,* Use of Natural resources for sustainable development, Edited by: Dr. Apurba Saikia, ISBN-978-93-82976-81-3, 2014, pp-9-11.

9. Dr. Durga Prasad Gogoi, *Diluted magnetic semiconductor nanostructures*, Material science and nanomaterials, recent advances and Applications, Edited by: Surajit konwer and Ankur Gogoi, Chaper-10, ISBN-978-93-81563-65-6, 2015, pp-163-183

10. Dr. Durga Prasad Gogoi, *Green nanotechnology for sustainable development*, Biodiversity: Conservation, Crisis & Sustainable use, Edited by Dr. Jonali Saikia Borkakoty, ISBN-978-81-930-653-2-7, 2015, pp-58-63,

11. Durga Prasad Gogoi, *Impacts of Semester System: A comparative case study with the Annual system*, Examination reforms and Evaluation in Semester System at Undergaraduate Level, Edited by: Anita Mahanta and Kunja Mukul Gogoi, ISBN-978-81-929944-4-4, 2016, pp-1-10.

12. Dr. Durga Prasad Gogoi, *Fabrication of ZnO and Mn:ZnO nanostructures and their Possible application in design of Biological marker*, IOSR Journal of Applied physics (IOSR-JAP), e-ISSN: 2278-4861, Vol 8, Issue 6 Ver. III, Nov-Dec, 2016, pp-30-33.

13. Durga Prasad Gogoi, *Exploring Physical and Optical Behavior of Co:ZnO Nanostructures*, International Journal of Innovations in Engineering and Technology (IJIET), ISSN-2319-1058, Vol-9, Issue 3, February 2018, pp-29-033.

14. Dr. Durga Prasad Gogoi, *Controlling student community at the age of internet*, Innovation and Rejuvenation of Teaching in Higher Education, Edited by: Nibedita phukan and Jayanta Sonowal, ISBN-978-93-82283-63-8, 2019, PP-138-144.

15. Durga Prasad Gogoi, *Fabrication of Co doped ZnO nanostructuresFabricated through solid state chemical reaction*, Ajanta,, ISSN-2277-5730, Vol-VIII, Issue-II, April-June-2019, pp-7-13.

References

1. Kellenberger, E. Assembly in Biological Systems, In: *Polymerization in Biological Systems*, CIBA Foundation Symposium, Amsterdam: Elsevier, 1972.

2. Drexler, K.E. *Engines of Creation*, Anchor Press, Doubleday, New York, 1986.

3. Eigler, D.M. & Schweizer, E.K., Positioning single atom with a scanning tunnelling microscope, *Nature (Lond.)* **344**, 524-526, 1990.

4. Ormos, P. Et al., Protein-based integrated optical switching and modulation, *Appl. Phys. Lett.* **80**, 2002.

5. Ramsden, J. What is nanotechnology, *Nanotechnology Perceptions* **1**, 3-17, 2005.

6. Bawendi, M.G. et al. The Quantum Mechanics of Larger Semiconductor Clusters ("Quantum Dots"), *Annu, Rev. Chem.* **41**, 477, 1990.

7. Millar, R.J.D. et al. *Surface electron transfer processes*, VCH: New York, 1995.

8. Alivisatos, A. P. Persepectives on the physical chemistry of semiconductor nanocrystals, *J. Phys. Chem.* **100**, 13226, 1996.

9. Murray, C.B. et al. Synthesis and Characterization of monodisper nanocrystals and closed packed nanocrystal assemblies, *Rev. Mater. Sc.*, **30**, 545, 2000.

10. Hu, Jiangatao et al. Chemistry and Physics in One Dimension: Synthesis and Properties of Nanowires and Nanotubes, *Acc. Chem. Res.* **32(5)**, 435-445, 1999.

11. Eychmüller, A., et al. Structure and Photophysics of Semiconductor Nanocrystals, *J. Phys. Chem. B* **104**, 6514, 2000.

12. Murphy C.J. & Coffer, J.L. Quantum Dots: A Primer, *Appl. Spectrosc.* **56**, 16A-27A, 2002.

13. Shim, M. et al. Doping and Charging in Colloidal Semiconductor anocrystals, *MRS Bull.* **26**, 1005, 2001.

14. Efros A.L. & Rosen, M. The electronic structure of semiconductor nanocrystals, *Annu. Rev. Mater. Sci.* **30**, 475, 2000.

15. Chan, W. C. W. & Nie, S. Quantum Dot Bioconjugates for Ultrasensitive Nonisotopic Detection, *Science* **281**, 2016, 1998.

16. Gao, X. & Nie, S. Molecular profiling of single cells and tissue specimens with quantum dots, *Trends Biotechnol.* **21**, 371, 2003.

17. Bruchez, M.M. et al. Semiconductor nanocrystals as fluorescent biological labels, *Science* **281**, 2013, 1998.

18. Zaban, A. et al. Photosensitization of Nanoporous TiO2 Electrodes with InP Quantum Dots, *Langmuir* **14**, 3153, 1998.

19. Plass, R. et al. Quantum Dot Sensitization of Organic-Inorganic Hybrid Solar Cells, *J. Phys. Chem. B* **106**, 7578, 2002.

20. Huynh, W.U. et al. Hybrid Nanorod-Polymer Solar Cells, *Science* **295**, 2425, 2002.

21. Klimov, V.I. et al. Optical gain and Stimulated Emission in Nanocrystal Quantum Dots, *Science* **290**, 314, 2000.

22. Colvin, V.L. et al. Light-Emitting-Diodes Made From Cadmium Selenoid Nanocrystals and a Semiconducting, *Nature (London)* **370**, 354, 1994.

23. Dabbousi, B.O. et al. Electroluminescence from CdSe quantum-dot/polymer composites, *Appl. Phys. Lett.* **66**, 1316, 1995.

24. Coe, S. et al. Electroluminescence from single monolayers of nanocrystals in molecular organic devices, *Nature (London)* **420**, 800, 2002.

25. Kastner, M.A. Artificial atom, *Phys. Today,* **46**, p.24, 1993.

26. Ekimov, A.I. & Onushchenko, A. A. Quantum size effect in the optical-spectra of semiconductor micro-crystals, *Sov. Phys. Semicond.* **16**, 775 1982.

27. Efros, A.L. & Efros, A.L. Interband absorption of light in a semiconductor sphere, *Sov. Phys. Semicond.* **16**, 772, 1982.

28. Papavassiliou, G.C. Luminescence spectra and Raman excitation profiles in small CdS particles, *J. Solid State Chem.* **40**, 330, 1981.

29. Brus, L.E. A simple model for ionization potential, electron effinity and aqueous redox potentials of small semiconductor crystallites, *J. Chem. Phys.* **79**, 5566-5571, 1983.

30. Brus, L.E. Electro-electron and electron-hole interactions in small semiconductor crystallites- the size dependant of the lowest excited electronic state, *J. Chem. Phys.* **80**, 4403, 1984.

31. Gaponenklo, S. V. *Optical Properties of Semiconductor Nanocrystals*, Cambridge University Press, USA, 1998.

32. Wang Y. & Herron N. Nanometer-Sized Semiconductor Clusters: Materials Synthesis, Quantum Size Effects, and Photophysical Properties *J. Phys. Chem.* **95,** 525, 1991.

33. Klimov, V., Haring Bolivar P. & Kurz, H. Ultrafast carrier dynamics in semiconductor quantum dots, *Phys. Rev. B.* **53**, 1463, 1996.

34. Bhargava, R.N. Doped nanocrystalline materials- Physics and applications, *J. Lumin.* **70,** 416, 1996.

35. Chen, W.C.W. & Nie, S. Quantum Dot Bioconjugates for Ultrasonic Nonisotopic Detection, *Science* **281,** 2016, 1998.

36. Parsapour, F., et al. Electron transfer dynamics in MoS_2 nanoclusters: Normal and inverted behaviour, *J. Chem. Phys.* **104,** 4978, 1996.

37. Waldrip, K.E. et al. Improved brightness, efficiency and stability of sputter deposited alternating current thin film electroluminescent ZnS:Mn by co-doping with potassium chloride, *Appl. Phys. Lett.* **76,** 1276, 2000.

38. Furdyna J.K. & Kossut, J. Semiconductors and Semimetals, R. Willardson and A. C. Beer, Eds., *Academic, New York*, 1988.

39. Furdyna, J.K. Diluted magnetic semiconductors, *J. Appl. Phys.* **64,** R29, 1988.

40. Bacher, G., et. al. Monitoring Statistical Magnetic Fluctuations on the Nanometer Scale, *Phys. Rev. Lett.* **89,** 127201, 2002.

41. Loss, D. et al. Quantum computation with quantum dots, *Phys. Rev. A* **57,** 120, 1998.

42. DiVincenzo, D. P., et al. Universal quantum computation with the exchange interaction, *Nature (London)* **408,** 339, 2000.

43. Leuenberger, M. N. & Loss, D. Quantum computing in molecular magnets, *Nature (London)* **410,** 789, 2001.

44. Kane, B.E. A silicon-based nuclear spin quantum computer, *Nature (London)* **393,** 133, 1998.

45. Afros, A.LO., et al. Paramagnetic Ion-Doped Nanocrystal as a Voltage-Controlled Spin Filter, *Phys. Rev. Lett.* **87,** 206601, 2001.

46. Xin, S.H., et al. Formation of self assembling CdSe quantum dots on ZnSe by molecular beam epitaxy, *Phys. Rev. Lett.* **69,** 3884, 1996.

47. Kim, C.S., et al. CdSe quantum dots in a $Zn_{1-x}Mn_xSe$ matrix: new effects due to the presence of Mn, *J. Crystal Growth* **214,** 395, 2000.

48. Maksimov A., et al. Magnetic polarons in a single diluted magnetic semiconductor quantum dot, *Phys. Rev. B* **62,** R7767–R7770, 2000.

49. Kossut, J., et al. Cathodoluminescence study of diluted magnetic semiconductor quantum well/micromagnet hybrid structures, *Appl. Phys. Lett.* **79,** 1789, 2001.

50. Prinz, G. A. Hybrid Ferromagnetic-Semiconductor Structure, *Science* **250,** 1092, 1990.

51. Tanaka, M. Epitaxial growth and properties of III–V magnetic semiconductor (GaMn) As and its heterostructures, *J. Vac. Sc. Technol.* **B16,** 2267, 1998.

52. Kamilla, S. K. & Basu, S. New semiconductor materials for magnetoelectronics at room temperature, Bull. Matter. Sc., 25, 6, 541-543, 2002.

53. Furdyna, J. K. Diluted magnetic semiconductors: Issues and opportunities *J. Vac. Sci. Technol. A* **4**, 2002, 1986.

54. Ohno, Y. et al. Electrical spin injection in a ferromagnetic semiconductor heterostructure, *Nature (London)* **402**, 790, 1999.

55. Prinz, G.A. Spin Polarized Transport, *Phys. Today* **48**, 58, 1995.

56. *Haury*, A., et al. Observation of a Ferromagnetic Transition induced by Two Dimensional Hole Gas in Modulated–Doped CdMnTe Quantum Wells, *Phys. Rev. Lett.* **79**, 511, 1997.

57. Ferrand, D. et al. Carrier-induced ferromagnetism in p-$Zn_{1-x}Mn_xTe$, *Phys. Rev. B* **63**, 085201, 2001.

58. Haury, A. et al. Observation of a Ferromagnetic Transition Induced by Two-Dimensional Hole Gas in Modulation-Doped CdMnTe Quantum Wells, *Phys. Rev. Lett.* **79**, 511, 1997.

59. Ferrand, D. et al. Carrier-induced ferromagnetism in p-$Zn_{1-x}Mn_xTe$, *Phys. Rev. B* **63**, 085201, 2001.

60. Matsumoto, H. et al. Preparation of Monodisperse CdSe nanocrystals by size selective photocorrosion, *J. Phys. Chem.* **100**, 13781, 2005.

61. Murray, C.B. et al., Synthesis and characterization of nearly monodisperse CdE (E= sulphur, selenium, tellurium) semiconductor nanocrystallites, *J. Am. Chem. Soc.* **115**, 8706, 1993.

62. Schmidt, G., et al., Fundamental obstacle for electrical spin injection from a ferromagnetic metal into a diffusive semiconductor *Phys. Rev. B* **62**, R4790, 2000.

63. Gaj, J. A., et al. Magneto-optical study of interface mixing in the CdTe-(Cd, Mn)Te system, *Phys. Rev. B* **50**, 5512, 1994.

64. Ossau, W.J., & Kuhn-Heinrich, B. Dimensional dependence of antiferromagnetism in diluted magnetic semiconductor heterostructures, *Physica B* **184**, 422, 1993.

65. Grieshaber, W., et al. Magneto-optic study of the interface in semimagnetic semiconductor heterostructures: Intrinsic effect and interface profile in CdTe-$Cd_{1-x}Mn_xTe$, *Phys. Rev. B* **53**, 4891, 1996.

66. Lee, S. & Dorbowolska, M. Enhancement of Zeeman splitting in double quantum wells containing ultrathin magnetic semiconductor layers, *Physica E* **10**, 300-304, 2001.

67. Ram-Mohan, L.R., et al. Transfer-matrix algorithm for the calculation of the band structure of semiconductor superlattices, *Phys. Rev. B* **38**, 6151, 1998.

68. Pedgeon, C.R. & Brown, R.N., Interband Magneto-Absorption and Faraday Rotation in InSb, *Phys. Rev.* **146**, 575, 1996.

69. Kane, E.O. Babd Structure of Indium Antimonide, *J. Phys. Chem. Solids* **1**, 249, 1973.

70. Lee, S., Dobrowolska, M. & Furdyna, J. K. Spin relaxation of excitons in nonmagnetic quantum dots: Effect of spin coupling to magnetic semiconductor quantum dots, *J. Appl. Phys.* **99**, 08F702, 2006.

71. Matskura, F., et al. Transport properties and origin of ferromagnetism in (Ga, Mn)As, *Phys. Rev. B* **57**, R2037, 1998.

72. Ohno, H. Making Nonmagnetic Semiconductors Ferromagnetic, *Science* **281**, 951, 1998.

73. Furdyna, J.K. Diluted Magnetic Semiconductors, *J. Appl. Phys. Rev.* **64**, R29, 1988.

74. Onodera, K., Masumoto, T., Kimura, M. 980 nm compact optical isolators using Cdl, Mn, Hg_2Te single crystals for high power pumping laser diodes, *Electron. Lett.* **30**, 1954, 1994.

75. Koshihara, S., et al. Ferromagnetic Order Induced by Photogenerated Carriers in Magnetic III-V Semiconductor Heterostructures of (In, Mn)As GaSb, *Phys. Rev. Lett.* **78**, 4617, 1997.

76. Ohno, H., et al. Electric-field control of ferromagnetism, *Nature* **408**, 944, 2000.

77. Wang, K.Y., et al. *Proceeding of the ICPS* **-26** (IOP, UK), 2002, p.58.

78. Edmonds, K. Y. et al. High-Curie-temperature $Ga_{1-x}Mn_xAs$ obtained by resistance-monitored annealing, *Appl. Phys. Lett.* **81**, 4991, 2002.

79. Hiba, D., et al. Effect of low-temperature annealing on (Ga, Mn)As trilayer structures, *Appl. Phys. Lett.* **82**, 3020, 2003.

80. Dietl, Tomasz, Spin order manipulations in nanostructures of II–VI ferromagnetic semiconductors, *J. of Magnetism and Magnetic Materials* **272-276** (Part 3) 1969-1973, 2004.

81. Egues, J. Carlos. Spin-Dependent Perpendicular Magnetotransport through a Tunable $ZnSe/Zn_{1-x}Mn_xSe$ Heterostructure: A Possible Spin Filter?, Physical Review Letters **80(20)**, 4578-4581, 1998.

82. Guo, Y. et al. Spin-polarized transport through a $ZnSe/Zn_{1-x}Mn_xSe$ heterostructure under an applied electric field, *J. Appl. Phy.* **88**, 6614, 2000.

83. Chang, K., et al., Spin polarized tunnelling through diluted magnetic semiconductor barriers, *Solid State Commun.* **120**, 181, 2001.

84. F. Zhai, et al., Effects of conduction band offset on spin-polarized transport through a semimagnetic semiconductor heterostructure, *J. Appl. Phys.* **90**, 1328, 2001.

85. Saffarzadeh, A. The effects of Mn concentration on spin-polarized transport through ZnSe/ZnMnSe/ZnSe heterostructures, *Solid state communication* **137**, 463-468, 2006.

86. Mac, W., et al. Ferromagnetic p-d exchange in $Zn_{1-x}Cr_xSe$ diluted magnetic semiconductor, *Phys. Rev. Lett.* **71**(1993)2327.

87. Mac, W., et al. *Proc.22nd Int. Conf. on the physics of semiconductors, Vancouver, Canada,* D.J. Lockwood (ed.), World Scientific, Singapore, 1995, p.2573.

88. Blinowski, J. and Kacman, P. Kinetic exchange in diluted magnetic semiconductors, ***Phys. Rev. B 46,*** 12298, 1992.

89. Blinowski, J. and Kacman, P. *Mater. Sci. Forum* 83-87(1992)523.

90. Blinowski, J. & Kacman, P. Ferromagnetism in Cr-based diluted magnetic semiconductors, *Journal of Crystal growth* **159**, 972-975, 1996.

91. Mac, W., Twardowski, A. & Demianiuk, M. *s, p-d* exchange interaction in Cr-based diluted magnetic semiconductors *Phys. Rev. B* **54,** 5528, 1996.

92. Wojtowicz, T., Karczewski, G. & Kossut, J. Excitons in novel diluted magnetic semiconductor quantum structures, *Thin Solid Films* **306,** 1997.

93. Saito, H., et al. Ferromagnetism in II–VI diluted magnetic semiconductor $Zn_{1-x}Cr_xTe$, *J. Appl. Phys.* **91,** 8085, 2002.

94. Saito, H., et al. Magneto-optical studies of ferromagnetism in the II-VI diluted magnetic semiconductor $Zn_{1-x}Cr_xTe$, ***Phys. Rev. B 66***, 081201(R), 2002.

95. Saito, H., et al. Room-Temperature Ferromagnetism in a II-VI Diluted Magnetic Semiconductor $Zn_{1-x}Cr_xTe$, *Phys. Rev. Lett.* **90**, 207202, 2003.

96. Mizokawa, T. et al. Electronic structure of the oxide-diluted magnetic semiconductor $Zn_{1-x}Mn_xO$, ***Phys. Rev. B 65,*** 085209, 2002.

97. Yichun, L., et al. Biocompatible ZnO/Au nanocomposites for ultrasensitive DNA detection using resonance Raman scattering, *J Phys. Chem. B* **112**, 6484, 2008.

98. Al-Hilli, S.M., et al. Zinc oxide nanorod for intercellular pH sensing, *Appl. Phys. Lett.* **89**, 173119, 2006.

99. Sadik, P.W., et al. Functionalizing Zn- and O-terminated ZnO with thiols, *Appl. Phys.* **101**, 104514, 2007.

100. Rao, C.N.R. & Deepak, F.L. Absence of ferromagnetism in Mn- and Co-doped ZnO, *J. Mater. Chem.* **15,** 573, 2005.

101. Spaldin, N.A. Search for ferromagnetism in transition-metal-doped piezoelectric ZnO, ***Phys.Rev. B 69,*** 125201, 2004.

102. Hong, N.H., et al. Observation of ferromagnetism at room temperature in ZnO thin films, *J. Phys. Condense. Matte,* **19,** 036219, 2007.

103. Kittilstved, K.R., et al. Direct Kinetic Correlation of Carriers and Ferromagnetism in Co^{2+}: ZnO, *Phys.Rev. Lett.* **97,** 037203, 2006.

104. Dietl, T., et al. Zener Model Description of Ferromagnetism in Zinc-Blende Magnetic Semiconductors, *Science* **287,** 1019, 2000.

105. Wimmer, E., et al., Full-potential self-consistent linearized-augmented-plane-wave method for calculating the electronic structure of molecules and surfaces: O_2 molecule, *Phys.Rev. B* **24**, 864, 1981.

106. Fukumura, T., et al. An oxide-diluted magnetic semiconductor: Mn-doped ZnO, *Appl. Phys.Lett.* **75**, 3366, 1999.

107. Jin, Z., et al. High throughput fabrication of transition-metal-doped epitaxial ZnO thin films: A series of oxide-diluted magnetic semiconductors and their properties, *Appl. Phys. Lett.* **78**, 3824, 2001.

108. Kim, J.-H., et al. Magnetic properties of epitaxially grown semiconducting Zn1ÀxCoxO thin films by pulsed laser deposition, *J.Appl. Phys.* **92**, 6066, 2002.

109. Tiwari, A., et al. Structural, optical and magnetic properties of diluted magnetic semiconducting $Zn_{1-x}Mn_xO$, *Solid State. Commun.* **121**, 371, 2002.

110. Chambers, S.A., et al. Epitaxial growth and properties of ferromagnetic co-doped TiO_2 anatase, *Appl. Phys. Lett.* **79**, 2001.

111. Park, W.K. Semiconducting and ferromagnetic behavior of sputtered Co-doped TiO2 thin films above room temperature, *J.Appl. Phys.* **91**, 8093, 2002.

112. Seong, N.-J., Yoon, S.-G. & Cho, C.-R. Effects of Co-doping level on the microstructural and ferromagnetic properties of liquid-delivery metalorganic-chemical-vapor-deposited $Ti1_{-x}Co_xO_2$ thin films, *Appl. Phys. Lett.*, **81**, 4209, 2002.

113. Venkatesan, M. et al., Anisotropic Ferromagnetism in Substituted Zinc Oxide, *Phys. Rev. Lett.*, 93, 177206, 2004.

114. Kim, D.H. et al. Formation of Co nanoclusters in epitaxial $Ti_{0.96}Co_{0.04}O_2$ thin films and their ferromagnetism, *Appl. Phys. Lett.*, **81**, 2421, 2002.

115. Stampe, P.A., et al. Investigation of the cobalt distribution in TiO_2:Co thin films, *J. Appl.Phys.*, **92**, 7114, 2002.

116. Kim, J.-Y., et al. Ferromagnetism Induced by Clustered Co in Co-Doped Anatase TiO_2 Thin Films, *Phys. Rev. Lett.*, **90**, 17401, 2003.

117. Kimura, H., et al. Rutile-type oxide-diluted magnetic semiconductor: Mn-doped SnO_2, *Appl.Phys. Lett.*, **80**, 94, 2002.

118. Andrearczyk, T., et al., *Proc. Of the 25th Int. Conf. On the Phys. Of Semiconductors, Part I and II*, 235 (2001)

119. Avrutin, V., et al. Optical and electrical properties of ZnMnO layers grown by peroxide MBE, *Superlattices and Microstructure*, **39**, 291-298, 2006.

120. Diaconu, M., et al. Room-temperature ferromagnetic Mn-alloyed ZnO films obtained by pulsed laser deposition, *J. of Magnetism and Magnetic Materials*, **307**, 212-221, 2006.

121. Hou, Deng-lu, et al., Magnetic properties of Mn-doped ZnO powder and thin films, *Material Science and Engineering, B* **138,** 184-188, 2007.

122. Liu, J.J., Yu, M.H. & Zhou, W.L. Well-aligned Mn-doped ZnO nanowires synthesized by a chemical vapour deposition method, *Appl. Phys. Lett.,* **87,** 172505, 2005.

123. Philipose, U., et al. High-temperature ferromagnetism in Mn-doped ZnO nanowires, *Appl. Phys. Lett.,* **88,** 263101, 2006.

124. Rubi, D., et al. Structural and magnetic properties of ZnO:TM (TM: Co, Mn) nanopowders, *Magnetism and Magnetic Materials* **316,** 2, 211-214, 2007.

125. Li, X.Z., Zhang, Jun & Sellmyer, D.J. Structural study of Mn-doped ZnO films by TEM, *Solid State Communication,* 141, 398-401, 2007.

126. Matsumutu, Y. et al. Room-Temperature Ferromagnetism in Transparent Transition Metal Doped Titanium Dioxide, *Science,* **291,** 854, 2001.

127. Ueda, K., Tabata, H. and Kawai, T. Magnetic and electric properties of transition-metal-doped ZnO films, *Appl.Phys. Lett.* **79,** 988, 2001.

128. Ogale, S.B. et al. High Temperature Ferromagnetism with a Giant Magnetic Moment in Transparent Co-doped $SnO_{2-\delta}$, ***Phys. Rev. Lett.*** **91,** 077205, 2003.

129. Shinde, S. R. and Ogale, S. B., et at. Ferromagnetism in laser deposited anatase $Ti_{1-x}Co_xO_{2-\delta}$ films, *Phys. Rev. B* **67,** 115211, 2003.

130. Cullity, B. D. *Introduction to Magnetic Materials,* Reading, Mass., Addison-Wesley Pub. Co., 1972.

131. Jungwirth, T., et al. Theory of ferromagnetic (III, Mn)V semiconductors, *Rev. Mod. Phys.* **78,** 809–864, 2006.

132. Nadeem Tahir, Surface Defects: Possible Source of Room Temperature Ferromagnetism in Co-Doped ZnO Nanorods, J. Phys. Chem. C 2013, 117, 8968–8973

133. Martín-González, M.S. et al., Insights into the room temperature magnetism of ZnO/Co_3O_4 mixtures. *J. Applied Physics,* **103,** 8, 083905 - 083905-4, 2009.

134. Abolfath, et al. Theory of magnetic anisotropy in $III_{1-x}Mn_xV$ ferromagnets, *Phys. Rev. B* **63,** 054418, 2001.

135. Kingery, W. D., et al. *Introduction to Ceramics,* Wiley, 1976.

136. Hu, J. T., et al. Linearly polarized emission from colloidal semiconductor quantum rod, *Science* **292,** 2060, 2001.

137. Schlam, E. Electroluminescent Phosphors, *Proc. IEEE* **61,** 894, 1973.

138. Lu W. & Lieber, C.M. Semiconductor nanowires, *J. Phys. D* **39,** R387, 2006.

139. Das, U., Mohanta D. & Choudhury, A. *Ind. J. Phys.* **81,** 155-159, 2007.

140. Datta, A. et al., Multicolor luminescence from transition metal ion (Mn^{2+} and Cu^{2+}) doped ZnS nanoparticles, J Nanoscience Nanotechnol, 7, 10, 3670-3676, 2007.

141. Wu, Y., et al. *Jpn. J. Appl.Phys.* **36**, LI 648, 1997.

142. Shibata, K., et al. Magneto-luminescence in $Cd_{1-x}Mn_xSe/ZnSe$ and $CdSe/Zn_{1-y}Mn_ySe$ quantum dots and quantum wells, *Physica E* **10**, 358, 2001.

143. Kim, C.S., et al. CdSe quantum dots in a $Zn_{1-x}Mn_xSe$ matrix:new e!ects due to the presence of Mn, *J. Cryst. Growth* **395**, 214-215, 2000.

144. Bhargava, R.N. and Gallagher, D. Optical properties of manganese-doped nanocrystals of ZnS, *Phys. Rev. Lett.* **72,** 416, 1994.

145. Bhargava, R.N. Doped nanocrystalline materials- Physics and applications, *J. Lumin.* **70,** 416, 1996.

146. SMITH, A. M., Engineering Luminescent Quantum Dots for In Vivo Molecular and Cellular Imaging, *Annals of Biomedical Engineering*, **34**, 1,3-14, 2006.

147. Parsapour, F., et al. Electron transfer dynamics in MoS_2 nanoclusters: Normal and inverted behaviour, *J. Chem. Phys.* **104**, 4978, 1996.

148. Waldrip, K.E., et al. Improved brightness, efficiency and stability of sputter deposited alternating current thin film electroluminescent ZnS:Mn by co-doping with potassium chloride, *Appl. Phys. Lett.* **76**, 1276, 2001.

149. Adachi, D., et al. Structural and luminescence properties of nanostructured ZnS:Mn, *Apply. Phys. Lett.* **77**, 1301, 2000.

150. Dinsmore, A.D., et al. Structure and luminescence of annealed nanoparticles of ZnS:Mn, *J.Appl. Phys.* **88**, 4985, 2000.

151. Bol, A.A., et al. Luminescence of nanocrystalline $ZnS:Cu^{2+}$, *J. Lumin.* **99**, 325, 2002.

152. Huang, J.M. et al. Photoluminescence and electroluminescence of ZnS:Cunanocrystals in polymeric networks, *Appl. Phys. Lett.* **70**, 2335, 1997.

153. Yang, P., et al. Luminescence of Cu^{2+} and In^{3+} co-activated ZnS nanoparticles, *Opt. Mater.* **20**, 141, 2002.

154. Kushida, T., et al. Optical properties of Sm-doped ZnS nanocrystals, *J. Lumin.* **87**, 466, 2000.

155. Kurita, A., et al. Wave length and angle-selective properties of optical memory effect by interference of multiple scattered light in Sm-doped ZnS nanocrystals, *J. Lumin.* **87**, 986, 2002.

156. Chen, W., et al. Upconversion luminescence of Eu^3 and Mn^2 in $ZnS:Mn^2$, Eu^3 codoped nanoparticles, *J. Appl. Phys.* **95**, 667, 2004.

157. Yang, P., et al. Photoluminescence characteristics of ZnS nanocrystallites co-doped with Co^{2+} and Cu^{2+}, Inorg. Chem. Commun. 4, 12, 734-737, 2001.

158. Z.L. Wang, Z. Zhang and Y. Liu, *Handbook of Nanophase and Nanostructured Materials-Synthesis*, Tsinghua University Press and Kluwer Academic/Plenum Publishers, 2003.

159. Xu, S.J., et al. Luminoscence characteristics of impurities-activated ZnS nanocrystals prepared in microemulsion with hydrothermal treatment,, *Appl. Phys. Lett.* **73**, 748, 1998.

160. Cao, L.X., et al. Luminescence enhancement of core-shell ZnS:Mn/ZnS nanoparticle, *Appl. Phys. Lett.* **80**, 4300, 2002.

161. Jiang, X., et al. Simultaneous In Situ Formation of ZnS Nanowires in a Liquid Crystal Template by γ-Irradiation, *Chem. Mater.* **13**, 1213, 2001.

162. Wang, Y., et al. Catalytic growth and photoluminescence properties of semiconductor single-crystal ZnS nanowires, *Chem. Phys. Lett.* **357**, 314, 2002.

163. Wang, X.D., et al. Rectangular Porous ZnO-Zns Nanotubes and ZnS Nanotubes, *Adv. Mater* **14**, 1732, 2002.

164. Zhu, Y.C., Bando, Y. and Uemura, Y. ZnS–Zn nanocables and ZnS nanotubes, *Chem. Commun.*, 836, 2003.

165. Moore, M. D., Li, J. & Zhong, W. L. Nanobelts, Nanocombs, and Nanowindmills of wurtzite ZnS, *Adv. Mater.* **15**, 228, 2003.

166. Jiang, Y., et al. Hydrogen-Assisted Thermal Evaporation Synthesis of ZnS Nanoribbons on a Large Scale, *Adv. Mater.*, **15,** 323, 2003.

167. Mahadevan, P. et al. Unusual Directional Dependence of Exchange Energies in GaAs Diluted with Mn: Is the RKKY Description Relevant?, Phys Rev. Lett. **93**, 177201, 2004.

168. R.K. Das et al. Theory of spin-EPR shift: Application to Pb1-xMnxTe, *Phys. Rev. B* **72**, 035216, 2005.

169. Furdyna, J. K. Diluted Magnetic Semiconductors, *J. Appl. Phys. Rev.* **64**, R29, 1988.

170. Sapra, S. et al. Influence of Quantum Confinement on the Electronic and Magnetic Properties of (Ga, Mn)As Diluted Magnetic Semiconductor, Nano Lett., 2 (6), 605 – 608, 2002.

171. Viswanatha, R. and Sarma, D.D., Study of the Growth of Capped ZnO Nanocrystals: A Route to Rational Synthesis, *Chem. Euro. J.* **12**, 180, 2006.

172. (a) Sapra, S., et al. Unraveling Internal Structures of Highly Luminescent PbSe Nanocrystallites Using Variable-Energy Synchrotron Radiation Photoelectron Spectroscopy, *J. Phys. Chem.* B **110** (31), 15244–15250, 2006; (b) Nanda, J. and Sarma, D. D. Photoemission spectroscopy of size selected zinc sulfide nanocrystallites, *J. Appl. Phys.* **90**, 2504, 2001.

173. Viswanatha, R., et. al. Synthesis and Characterization of Mn-Doped ZnO Nanocrystals, *J. Phys. Chem. B*, 108 (20), 6303 – 6310, 2004.

174. Sharma, D.D., et. Al. Magnetic Properties of Doped II–VI Semiconductor Nanocrystals, *J. Nanosci. Nanotech*, **5**, 1503 – 1505, 2005.

175. Sapra, S., et. al. Emission Properties of Manganese-Doped ZnS Nanocrystals, *J. Phys. Chem.* B **109 (5)**, 1663 – 1668, 2005.

176. Venkataprasad Bhat, S. and Deepak, F.L. Tuning the bandgap of ZnO by substitution with Mn^{2+}, Co^{2+} and Ni^{2+}, *Solid State Communicatios.* **135**, 345-347, 2005.

177. Rao, C.N.R. and Deepak, F.L. Absence of ferromagnetism in Mn- and Co-doped ZnO, *J. Mater. Chem.* **15**, 573-578, 2005.

178. Paul D. et al. Synthesis of $Zn_{0.95}Cr_{0.05}O$ DMS by co-precipitation and ceramic methods: Structural and magnetization studies, *Materials Chemistry and Physics* **97**(2006) 188-192.

179. Deka, S. and Joy, P.A. Synthesis and magnetic properties of Mn doped ZnO nanowires, *Solid state Communication* **142**, 190-194, 2007.

180. Karar, N., Raj, S. and Singh, F. Properties of nanocrystalline ZnS:Mn, *Journal of Crystal Growth* **268**, 585-589, 2004.

181. Mohanta, D. et al. Properties of 80-MeV oxygen ion irradiated ZnS:Mn nanoparticles and exploitation in nanophotonics, *J. of Nanoparticle Research,* **8,** 645-652, 2006.

182. Mohamta D. et al. Irradiation induced grain growth and surface emission enhancement of chemically tailored ZnS:Mn/PVOH nanoparticles by Cl^{+9} ion impact, *Bull. Matter. Sci.* **26**(3), 289-294, 2003.

183. Baibich, M. et al. Giant Magnetoresistance of (001)Fe/(001)Cr Magnetic Superlattices, *Phys. Rev. Lett.* **61**, 2472, 1988.

184. Binash, G. et al. Enhanced Magnetoresistance in Layered Magnetic Structures with Antiferromagnetic Interlayer Exchange, *Phys. Rev. B.* **39**, 4828, 1989.

185. Parkin, S. S. P., Li, Z. G. and Smith, D. J. Giant Magnetoresistance in Antiferromagnetic Co/Cu Multilayers, *Appl. Phys. Lett.* **58**, 2710, 1991.

186. Dieny, B. et al., Giant Magnetoresistance in Soft Ferromagnetic Multilayers, *Phys. Rev.* B **43**, 1297, 1991.

187. MRAM Basics. Weblink: http://www.crocus-technology.com/technology.html

188. Datta & Das. Electronic analog of the electro-optic modulator, *Appl. Phys. Lett.* **56**, 665, 1990.

189. Zutíc, I. Fabian, J., Sarma, S. D. Spintronics: Fundamentals and applications, *Rev. Mod. Phys.* **76**, 323–410, 2004.

190. Mohanta, D. et al. Properties of 80-MeV oxygen ion irradiated ZnS:Mn nanoparticles and exploitation in nanophotonics, *J. of Nanoparticle Research,* **8,** 645-652, 2006.

191. Gamelin et al. Synthesis of colloidal Mn^{2+}:ZnO quantum Dots and High-Tc ferromagnetic Nanocrystalline Thin films, *J. Am. Chem. Soc.* **126**, 9387-9398, 2003.

192. Wang, Z. L., Zhang, Z. and Liu, Y. (editors). *Handbook of nanophase and Nanostructured Materials-Synthesis*, Tsnghua University Press and Kluwer Academic/Plenum Publishers, 2003.

193. Wu, S.H. and. Chen, D.H. Synthesis of high-concentration Cu nanoparticles in aqueous CTAB solutions, *J. Colloid and Interface Sci.* **273**(1), 165–169, 2004.

194. Li, F., Vipulanandan, C. and Mohanty, K.K. Microemulsion and solution approaches to nanoparticle iron production for degradation of trichloroethylene, *Colloids and Surfaces A, Physicochemical and Engineering Aspects*, **223(1-3)**, 103 – 112, 2003.

195. Jia, Y. et al. Hydroxyapatite nanostructure material derived using cationic surfactant as a template, *J. Materials Chemistry*, **13**(12), 3053 – 3057, 2003.

196. Lim, Y.Y. and Lim, K.H. Interaction of Aqueous Solutions of a Surface Active Copper(II) Complex with Several Common Surfactants, *J. Colloid and Interface Sci.* **196(1)**, 116 – 119, 1997.

197. Vittal, R., Gomathi, H. & Rao, G.P. Influence of a cationic surfactant on the modification of electrodes with nickel hexacyanoferrate surface films, *Electrochimica Acta* **45(13)**, 2083 – 2093, 2000.

198. Soror, T.Y. and. El-Ziady, M.A. Effect of cetyl trimethyl ammonium bromide on the corrosion of carbon steel in acids, *Mater. Chem. Phys.* **77(12)**, (2003), 697 – 703, 2003.

199. Kittel, C. Introduction to solid state physics, Willey, 2004.

200. Scherrer, P. Bestimmung der Grösse und der inneren Struktur von Kolloidteilchen mittels Röntgenstrahlen, *Nachr. Ges. Wiss. Göttingen* **26**, 98-100, 1918.

201. Skoog, D. A., West D. M., Holler, F. J. *Analytical Chemistry: An Introduction*, Saunders Golden Sunburst Series (7th Ed.), **1999**.

202. Brus, L. Electronic wave functions in semiconductor clusters: experiment and theory, *The Journal of Physical Chemistry* **90(12)**, 2555 – 2560, 1986.

203. Murphy, C.J. Optical Sensing with Quantum Dots, *Anal. Chem.* **74**, 520A-526A, 2002.

204. Hanneewald, K., Glutsch, S. and Bechstedt, F. Nonequilibrium theory of photoluminescence excitation spectroscopy in semiconductors, Phys. *Stat. Sol. (b)* **238(3)**, 517– 520, 2003.

205. Voutou, B. Stefanaki, E-C., Electron Microscopy: The Basics - Mansic, www.mansic. eu/documents/PAM1/Giannakopoulos1.pdf, 2013.

206. Yan, K. et al. Photoluminescence lifetime of nanocrystalline ZnS:Mn^{2+}, *Phys. Rev. B* **58**, 13585–13589, 1998.

207. Meyer, E. Atomic force microscopy, *Progress in Surface Science*, **41**, 3-49 1992.

208. SQUID, http://en.wikipedia.org/wiki/SQUID, 2013.

209. Ran, Shannon K'doah, *Gravity Probe B: Exploring Einstein's Universe with Gyroscopes*. http://einstein.stanford.edu/content/education/GP-B_T-Guide4-2008.pdf, 2013.

210. Drung, D. et al. Highly sensitive and easy-to-use SQUID sensors, *IEEE Transactions on Applied Superconductivity* **17** (2), 699, 2007.

211. Pan, Z. W., Dai, Z. R. and Wang, Z. L.. Nanobelts of Semiconducting Oxides, **Science** **291**, 1947, 2001.

212. Huang, M. H., et al., Room-Temperature Ultraviolet Nanowire Nanolasers, *Science* **292**, 1897, 2001.

213. Shannon, R.D., Revised effective ionic radii and systematic studies of interatomic distances in halides and chalcogenides, *Acta Cryst. A* **32**, 751, 1976.

214. Fukumura, T. et al. An oxide-diluted magnetic semiconductor: Mn-doped ZnO, *Appl. Phys.Lett.***75**, 3366, 1999.

215. Jung, S.W., et al. Ferromagnetic properties of $Zn_{1-x}Mn_xO$ thin films, *Appl. Phys. Lett.* **80**, 4561, 2002.

216. Tiwari, A., et al. Structural optical and magnetic properties of diluted magnetic semiconducting $Zn_{1-x}Mn_xO$ films, *Solid State Commun.* **121** 371, 2002.

217. Cullity, B.D. *Elements of X-Ray Diffraction*, 2nd ed., Addison-Wesley, Massachusetts, 1978.

218. J. Luo, J., et al. Structure and magnetic properties of Mn-doped ZnO nanoparticles, *J. Appl. Phys.* **97**, 086106, 2005.

219. Deka, S., Joy, P.A. synthesis and magnetic properties of Mn doped ZnO nanowires, *Solid State Commun.* **142**, 190–194, 2007.

220. Zang, J. F., et al. Tailoring zinc oxide nanowires for high performance amperometric glucose sensor, *Electroanal.* **19**, 1008–1014, 2007.

221. Baruah, S.; Dutta, J. p^H-dependent growth of zinc oxide nanorods, *J. Cryst. Growth*, **311**, 2549–2554, 2009.

222. Xu, S., et al. Growth and transfer of monolithic horizontal ZnO nanowire superstructures onto flexible substrates. *Adv. Funct. Mater.* **20**, 1493–1495, 2010

223. Laudise, R. A.; Ballman, A. A. Hydrothermal synthesis of zinc oxide and zinc sulphide, *Phys. Chem.* **64**, 688–691, 1960.

224. Xu, S. & Lin Wang, Z. One-Dimensional ZnO Nanostructures: Solution Growth and Functional Properties, *Nano Res.* **4(11)**, 1013-1098, 2011.

225. Banno, M., et al. Vibrational dynamics of hydrogen-bonded complexes in solutions studied with ultrafast infrared pump–probe spectroscopy, *Acc. Chem. Res.* **42(9)**, 1259–1269, 2009.

226. Crupi, V., et al. Rayleigh wing and Fourier transform infrared studies of intermolecular and intramolecular hydrogen bonds in liquid ethylene glycol, *Molec. Phys.* 84(4), 64–52, 1995.

227. Nyquist, R. A., et al. NMR and IR spectra–structure correlations for carbonyl containing compounds in various solvents, *J. Molec. Struct.* **377(2)**, 113–128, 1996.

228. Nyquist, R.A.; Putzig, C.L. et al. Solvent and concentration effects on carbonyl stretching frequencies, *Ketones Appl. Spectrosc.* 43(6), 983–991, 1989.

229. Coleman, M.M.; Lee, K.H. et al. Hydrogen bonding in polymers. 4.Infrared Temperature studies of a simple polyurethane, *Macromolecules* 19(8), 2149–2157, 1986.

230. Choperena, A. & Painter, P.C. Hydrogen bonding in polymers: effect of temperature on the OH stretching bands of poly(vinylphenol), *Macromolecules* **42(16)**, 6159–65, 2009.

231. Huang, H., et al. Two-dimensional correlation infrared spectroscopic study of n-methylacetamide as a function of temperature, *J. Phys. Chem.* A **107(39)**, 7697–7703, 2003.

232. Cao, B.Q., et al. Phosphorus acceptor doped ZnO nanowires prepared by pulsed-laser deposition, *Nanotechnology* **18**, 455707, 2007.

233. Khranovskyy, V., et al. Comparative PL study of individual ZnO nanorods, grown by APMOCVD and CBD techniques, *Physica, B. Condensed matter* **407**(10), 1538-1542, 2012.

234. Vanheusden, K., et al. Correlation between photoluminescence and oxygen vacancies in ZnO phosphors, *Appl. Phys. Lett.* **68**, 403, 1996.

235. Leiter, F., et al. Magnetic resonance experiments on the green emission in undoped ZnO crystals, *Physica B* **908**, 308, 2001.

236. Look, D.C., et al. Residual Native Shallow Donor in ZnO, *Phys. Rev. Lett.* **82**, 2552, 1999.

237. Wu, X. L., et al. Photoluminescence and cathodoluminescence studies of stoichiometric and oxygen deficient ZnO films, *Appl. Phys. Lett.* **78**, 2285, 2001.

238. Lin, B. X., Fu, Z. X. and Jia, Y. B. Green luminescent centre in undoped ZnO films deposited on silicon substrate, *Appl. Phys. Lett.* **79**, 943, 2001.

239. Bohle, D.S. & Spina, C.J. Nanocrystalline Zinc Oxide: Surface Influence on Photoluminescence and Photocatalysis. *J. Am. Chem. Soc.*, **131**, 4397-4404, **2009.**

240. Kim, S. Y., et al. Synthesis and Characterization of Novel Red-Light-Emitting Materials with Push-Pull Structure Based on Benzo[1,2,5] thiadiazole Containing Arylamine as an Electron Donor and Cyanide as an Electron Acceptor, *Bull. Korean Chem. Soc.* **29**, 1960, 2008.

241. Joo, J., et al. Synthesis of ZnO Nanocrystals with Cone, Hexagonal Cone, and Rod Shapes via Non-Hydrolytic Ester Elimination Sol–Gel Reactions, *Adv. Mater.* **17**, 1873, 2005.

242. Xu, P.S. et al. The electronic structure and spectral properties of ZnO and its defects, Nuclear Instruments and Methods in Physics Research Section B: Beam Interactions with Materials and Atoms, **199**, 286–290, 2003.

243. Aleksandra, B. D. & Leung, Y. H. Optical Properties of ZnO Nanostructures, *Small* **2**, 944, 2006.

244. Spanhel, L. & Anderson, A. Semiconductor clusters in the sol-gel process: quantized aggregation, gelation, and crystal growth in concentrated zinc oxide colloids, *J. Am. Chem. Soc.* **113(8)**, 2826–2833, 1991.

245. Zeng, H.B., et al., Violet photoluminescence from shell layer of Zn/ZnO core-shell nanoparticles induced by laser ablation, *Appl. Phys. Lett.* **88**, 171910, 2006.

246. Zhang, Y. et al. Selective Growth of Vertical ZnO Nanowire Arrays Using Chemically Anchored Gold Nanoparticles, *Appl. Phys. Lett.* **83**, 4631–4633, 2003.

247. Huang, M. H. et al. Catalytic Growth of Zinc Oxide Nanowires by Vapor Transport, *Adv. Mater.* **13**, 113, 2001.

248. Mahmood, K., Bin Park, S. and Jin Sung, H. Enhanced photoluminescence, Raman spectra and field emission behavior of indium-doped ZnO nanostructures, *J. Mater. Chem. C* **1**, 3138-3149, 2013.

249. Vanheusden, K. et al., Mechanisms behind green photoluminescence in ZnO phosphor powders, *J. Appl. Phys.* **79**, 7983, 1996.

250. Li, Y. et al. Ordered semiconductor ZnO nanowire arrays and their photoluminescence Properties, *Appl. Phys. Lett.* **76**, 2000, 2011.

251. Hu, J. Q. & Bando, Y. Growth and optical properties of single-crystal tubular ZnO whiskers, *Appl. Phys. Lett.* **82**, 1401, 2003.

252. Z. Peng, Z. et al. Photoluminescence and Raman analysis of novel ZnO tetrapod and multipod nanostructures, *Appl. Surf. Sci.* **256**, 6814–6818, 2010.

253. Zhang, H., Shen, L. and Guo, S. W. Self-catalytic Synthesis of ZnO Tetrapods, Nanotetraspikes, and Nanowires in Air at Atmospheric Pressure, *J. Phys. Chem. C* **111**, 12939, 2011.

254. Sernelius, B.E. et al. Band-gap tailoring of ZnO by means of heavy Al doping, *Phys. Rev. B: Condens. Matter* **37**, 10244 -10248, 1988.

255. Sanon, G., Rup, R. and Mansingh, A. Band-gap narrowing and band structure in degenerate tin oxide (SnO_2) films. *Phys. Rev. B: Condens.Matte* **44**, 5672 – 5680, 1991.

256. A. Wander, F. Schedin, P. Steadman, A. Norris, R. McGrath, T. S. Turner, G. Thornton and N. M. Harrison, Stability of Polar Oxide Surfaces, *Phys. Rev.Lett.* **86**, 3811–3814, 2001.

257. Fang, Y., Wong, K. M. and Lei, Y., Synthesis and field emission properties of different ZnO nanostructure arrays, ***Nanoscale Res. Lett.*** **7(1)**, 197, 2012.

258. Pal, B. and Giri, P. K. Defect Mediated Magnetic Interaction and High Tc Ferromagnetism in Co Doped ZnO Nanoparticles, *Journal of Nanoscience and Nanotechnology* **11**, 1–8, 2011.

259. Cui, J and Gibson, U. Thermal modification of magnetism in cobalt-doped ZnO nanowires grown at low temperatures, *Phys. Rev. B* **74** 045416, 2006.

260. Luna-Arredondo, E.J. et al. Indium-doped ZnO thin films deposited by the sol–gel technique, *Thin Solid Films* **490**, 132, 2005.

261. Majeed Khan, M.A. et al. Influences of Co doping on the structural and optical properties of ZnO nanostructured, *Appl. Phys. A* **100**, 45–51, 2010.

262. P. K. Giri, P.K. et al. Correlation between microstructure and optical properties of ZnO nanoparticles synthesized by ball milling *J. Appl. Phys.* **102**, 093515, 2007.

263. Z. W. Zhao, Z.W., et al. Large magnetic moment observed in Co-doped ZnO nanocluster assembled thin films at room temperature, *Appl. Phys. Lett.* **90**, 152502, 2007.

264. Ahmed, F. Et al., Doping effects of Co2+ ions on structural and magnetic properties of ZnO nanoparticles, *Microelectronic Engineering* **89**, 129-132, 2012.

265. Ullah R. and Dutta J. Photocatalytic degradation of organic dyes with manganese-doped ZnO nanoparticles, *Journal of Hazardous Materials* **156**, 194-200, 2008.

266. Sankara Reddy B., et al. Synthesis, Structural, Optical Properties and Antibacterial activity of codoped (Ag, Co) ZnO Nanoparticles, *J. Res. J. Material Sci.* **1(1)**, 11-20, 2013.

267. Wang, Y. X. et al., Properties of Co-doped ZnO films prepared by electrochemical deposition, *Cryst. Res. Technol.* **44(5)**, 517 – 520, 2009.

268. Sato, K. & Katayama-Yoshida, H., Material Design for Transparent ZnO-Based Magnetic Semiconductors, *Jpn. J. Appl. Phys.* **39**, L555, 2000.

269. Zhao, Z. W. et al. Large magnetic moment observed in Co-doped ZnO nanocluster-assembled thin films at room temperature, *Applied Physics Letters* **90,** 15, 152502 – 152502, 2007.

270. Kim, K. K., et al. Fabrication of ZnO quantum dots embedded in an amorphous oxide layer, *Appl. Phys. Lett.* **84**, 3810, 2004.

271. Fochs, P. D. The measurement of the energy gap of semiconductors from their diffuse reflection spectra, *Proceedings of the Physical Society. Section B* **69**(1), 70, 1955.

272. Joseph, D. Paul & Venkateswaran, C. Bandgap Engineering in ZnO By Doping with 3d Transition Metal Ions, *Molecular and Optical Physics* **2011**, 2011.

273. Tiwari, A., et al. Integration of single crystal $La_{0.7}Sr_{0.3}MnO_3$ films with Si(001), *Solid State Communications* **121**(6-7), 371–374, 2002.

274. Wójcik, A., et al. Monocrystalline and Polycrystalline ZnO and ZnMnO Films Grown by Atomic Layer Epitaxy Growth and Characterization, *Acta Physica Polonica A* **105**(6), 667–673, 2004.

275. Kane, M.H., et al. Surface chemistry of atomic layer deposition: A case study for the trimethylaluminum/water process, *Journal of Applied Physics* **97**(2), 023906, 2005.

276. Yin, Z., et al. Structural, magnetic properties and photoemission study of Ni-doped ZnO, *Solid Stat. Commun.* **135**, 430 - 433, 2005.

277. Wang, X.B., et al. The influence of different doping elements on microstructure, piezoelectric coefficient and resistivity of sputtered ZnO film, *Appl. Surf. Sci.* **253**, 1639 - 1643, 2006.

278. Cheng, C.W., Xu, G.Y., Zhang, H.Q., Luo, Y. Hydrothermal synthesis Ni-doped ZnO nanorods with room-temperature ferromagnetism, *Mater. Lett.* **62**, 1617, 2008.

279. S. Ghosh, S., et al. Study of ZnO and Ni-doped ZnO synthesized by atom beam sputtering technique, *Appl. Phys. A* **90**, 765 - 769, 2008.

280. Cong, C.J., et al. Synthesis, structure and ferromagnetic properties of Ni-doped ZnO nanoparticles, *Solid State Commun.* **138**, 511 – 515, 2006.

281. Wu, D., et al. Preparation and properties of Ni-doped ZnO rod arrays from aqueous solution, *J. Colloid Interface Sci.* **330**, 380 – 385, 2009.

282. Xu, P.S., et al. The electronic structure and spectral properties of ZnO and its defects, Nuclear instruments and methods, In: *Physics research section B: Beam interactions with materials and Atoms*, **199**, 286, 2003.

283. Klingshrin, C. F. *Semiconductor Optics*, Springer, Berlin, 1997.

284. Brus. L.E., et al. *Special Issue on Spectroscopy of Isolated and Assembled Semiconductor Nanocrystals*, *J. Lumin.*, Elsevier Science, 1996.

285. Toyama, T., et al. Thin-film electroluminescence device utilizing ZnS:Mn nanocrystals as emission layer, *J. Non-Cryst. Solids* **299**, 1111, 2002.

286. Behboudnia, M. & Sen, P. Systematics in the nanoparticle band gap of ZnS and $Zn_{1-x}M_xS$ (M=Mn, Fe, Ni) for various dopant concentrations, *Phys. Rev. B* **63**, 035316, 2001.

287. Bol, A. A., et al. On the incorporation of Trivalent rare Earth ions in II-VI Semiconductor Nanocrystals, *Chem. Mater* **14,** 1121, 2002.

288. Vacassy, R. et al., *Mater. Res. Soc. Symp. Proc.* **501,** 369, 1998.

289. Sreeekantha Reddy, D., Synthesis and characterization of $Zn_{1-x}Mn_xS$ nanocrystalline films prepared on glass substrates, *Appl. Phys. A***91,** 627, 2008.

290. Tiwari, A., et al. Structural, optical and magnetic properties of diluted magnetic semiconducting $Zn_{1-x}Mn_xO$ films, *Solid State Commun.***121,** 371, 2002.

291. Calandra, P., Goffred, M. and Livery, V. T. *Colloids Surf. A: Physicochemical and Engineering Aspects*, **160,** 9, 1999.

292. D Mohanta, P Saikia and A Chowdhury Indian J. Phys. **79(9)** 1015 (2005)

293. Senithkumar, S. and Selvi, R.T. Formation and photoluminescence of ZnS nanorods, *J. Appl. Sc.* **8,** 2306, 2008.

294. Mandelbrot, B. *The Fractal Geometry of Nature*, 3rd ed., W.H. Freeman and Company, New York, 1983.

295. Mandelbrot, B. *Fractal and Chaos, The Mandelbrot Set and Beyond*; Springer-Verlag, New York, 2004.

296. Lomander, A., Hwang, W., and Zhang, S. Hierarchical Self-Assembly of a Coiled-Coil Peptide into Fractal Structure, *Nano Lett.* **5,** 1255-1260, 2005.

297. Beck, C. & Schögl, F. *Thermodynamics of Chaotic Systems*, Cambridge University Press, Cambridge, 1993.

298. Skoog, D.; Leary, J. *Principles of Instrumental Analysis*, 4th ed., Saunders College Publishing, Philadelphia, 1992.

299. Meakin, P. Diffusion-controlled cluster formation in 2—6-dimensional space, *Phys. Rev. A* **27(3)**, 1495-1507, 1983.

300. Witten, T.; Sander, L. Diffusion-Limited Aggregation, a Kinetic Critical Phenomenon, *Phys. ReV. Lett.* **47**, 1400-1403, 1981.

301. Witten, T.; Sander, L. Diffusion-limited aggregation, *Phys. ReV. B* **27(9)**, 5686-5697, 1983.

302. Whitesides, G. M. The 'right' size in nanotechnology, *Nat. Biotechnol.* **21**, 1161-1165, 2003.

303. Zhang, S. Fabrication of novel biomaterials through molecular self-assembly, *Nat. Biotechnol.* **21**, 1171-1178, 2003.

304. Reches, M. & Gazit, E. Casting Metal Nanowires Within Discrete Self-Assembled Peptide Nanotubes, *Science* **300**, 625-627, 2003.

305. Kroger,N.,Deutzmann,R.,Sumper,M.Silica-precipitatingPeptidesfromDiatomsTHE CHEMICAL STRUCTURE OF SILAFFIN-1A FROM CYLINDROTHECA FUSIFORMIS, *J. Biol. Chem.* **276**, 26066-26070, 2001

306. Bratt, L. L.; Stone, M. O. Ultrafast holographic nanopatterning of biocatalytically formed silica, *Nature* **413**, 291-293, 2001.

307. Mann, S., Cölfen, H. Higher-Order Organization by Mesoscale Self-Assembly and Transformation of Hybrid Nanostructures, *Angew. Chem., Int. Ed.*, **42**, 2350-2365, 2003.

308. Gupta, P.; Rizwan, H., Khan, H.; Saleemuddin, M. Induction of 'molten globule' like state in acid denatured state of unmodified preparation of stem bromelain: Implication of disulphides in protein folding, *Int. J. Biol., Macromol.* **33(4-5)**, 167-174, 2003.

309. Horng,J.; Demarest, S.; Rsleigh, D. pH-dependent stability of the human α-lactabumin molten globule state: Contrasting roles of the 6–120 disulfide and the β-subdomain at low and neutral pH, *Proteins: Struct., Funct., Genet.*, **52**, 193-202, 2003.

310. Fukumura, T. et al. Magnetic properties of Mn-doped ZnO, *Appl. Phys. Lett.* **78(7)**, 958, 2001.

311. J. Luo, J. et al. Structure and magnetic properties of Mn-doped ZnO nanoparticles, *J. Appl. Phys.* **97**, 086106, 2005.

312. Chen, W., et al. Magnetism in Mn-doped ZnO bulk samples, *Solid State Commun.* **134**, 827, 2005.

313. Zhang, H.-W., et al. Magnetism in $Zn_{1-x}Mn_xO$ crystal prepared by hydrothermal method, *Solid State Commun.* **137**, 272, 2006.

314. ZHOU Shao-Min et al. Ferromagnetism from Co-Doped ZnO Nanocantilevers above Room Temperature, *Chin. Phy. Lett.* **25(12)**, 4446, 2008.

315. J. Spalek, J. Et al. Magnetic susceptibility of semimagnetic semiconductors: The high-temperature regime and the role of superexchange, *Phys. Rev. B*, **33**, 3407, 1986.

316. Dietl, T., et al. Zener Model Description of Ferromagnetism in Zinc-Blende Magnetic Semiconductors, *Science* **287**, 1019, 2000.

317. Matsumoto, Y. et al. Room-Temperature Ferromagnetism in Transparent Transition Metal-Doped Titanium Dioxide, *Science* **291**, 854, 2000.

318. Theodoropoulou, N., et al. Magnetic and structural properties of Mn-implanted GaN, *Appl. Phys. Lett.* **78**, 3475, 2001.

319. Cui, J., and Gibson, U., Thermal modification of magnetism in cobalt-doped ZnO nanowires grown at low temperatures, *Phys. Rev. B* **74**, 045416, 2006.

320. Schwartz, D. A., Kittilstved, K. R. and Gamelin, D. R. Above-room-temperature ferromagnetic Ni^{2+}-doped ZnO thin films prepared from colloidal diluted magnetic semiconductor quantum dots, *Appl. Phys. Lett.* *85*, 1395-1397, 2004.

321. Yan, H., et al. Above-room-temperature ferromagnetic Ni2+-doped ZnO thin films prepared from colloidal diluted magnetic semiconductor quantum dots, *Appl. Phys. Lett.* **90**, 082503, 2007.

322. Whitney, T. M. et al. Fabrication and Magnetic Properties of Arrays of Metallic Nanowires, *Science* **261**, 1316 – 1319, 1993.

323. Bødker, F., et al. Particle interaction effects in antiferromagnetic NiO nanoparticles, *J. Magn. Magn. Mater.* **221,** 32, 2000.

324. Liu, X.X., et al. Doping concentration dependence of room-temperature ferromagnetism for Ni-doped ZnO thin films prepared by pulsed-laser deposition, *Appl. Phys. Lett.* **88** 062508, 2006.

325. Singh S, et al. Correlation between electrical transport, optical, and magnetic properties of transition metal ion doped ZnO, *J. Appl. Phys.* **103**, 07D108, 2008.

326. Mao, X. Y., Zhong, W. & Du, Y. W. Ferromagnetism of Ni cluster in Ni-doped ZnO by solid state reaction, *J. Magn. Magn. Mater.* **320,** 1102, 2008.

327. Parra Palomino, A G. *Room-Temperature Synthesis and Characterization of Highly Monodisperse Transition Metal-Doped ZnO Nanocrystals*, Thesis, UNIVERSITY OF PUERTO RICO MAYAGÜEZ CAMPUS, 2006.

328. Liu, X.J. et al. Intrinsic and extrinsic origins of room temperature ferromagnetism in Ni-doped ZnO films, *J. Phys. D: Appl. Phys.* **42,** 035004, 2009.

329. Norton, D.P., et al. Ferromagnetism in Mn-implanted ZnO:Sn single crystals, *Appl. Phys. Lett.* **82**, 239, 2003.

330. Ohno, Y. Electrical spin injection in a ferromagnetic semiconductor heterostructure, *Nature,* **402**, 1999.

331. Holub, M. and Bhattacharya, P. Spin-polarized light-emitting diodes and lasers, *J. Phys. D: Appl. Phys.* **40,** 179–R203, 2007.

332. Dietl, T., Ohno, H. and Matsukura, F. Hole-mediated ferromagnetism in tetrahedrally coordinated semiconductors, *Phys. Rev. B* **63,** 195205, 2001.

333. Sivalingam, D. et al., Ethanol and trimethyl amine sensing by Zno-based nanostructured thin films. *Int. J. Nanosci.,* **10**, 1161–1165, 2011.

334. Sivalingam, D.; Gopalakrishnan, J.; Rayappan, J.B. Influence of precursor concentration on structural, morphological and electrical properties of spray deposited ZnO thin films. *Crystal Res. Technol.,* **46**, 685–690, 2011.

335. Sivalingam, D.; Gopalakrishnan, D. & Rayappan, R. Structural, morphological, electrical and vapour sensing properties of Mn doped nanostructured ZnO thin films. *Sens. Actuators B Chem.*, 166–167, 624–631, 2012.

336. artin Längkvist et al., Fast Classification of Meat Spoilage Markers Using Nanostructured ZnO Thin Films and Unsupervised Feature Learning, *Sensors,* 13, 1578-1592, 2013.

337. Abouelkaram, S. et al., Effects of muscle texture on ultrasonic measurements. *Food Chem.*, 69, 447–455, 2000.

338. Ophir, J. et al., Elastography of beef muscle. Meat *Sci.* 36, 239–250, 1994.